Collaborative Working in Construction

Complications arising from poor collaboration are the source of a variety of the construction industry's biggest problems. It is now widely recognised that an effective collaboration strategy based on the implementation of information systems and careful consideration of the wider organisational issues is the key to delivering construction projects successfully.

Against a backdrop of rapidly developing communication technologies, and continuing efforts to improve working practices, this book provides clear explanations of how to successfully devise and implement a collaboration strategy.

The concepts introduced include:

- collaborative working as a holistic concept in construction;
- a new framework on how to plan and implement effective collaboration;
- change management approaches for introducing collaborative working systems, and implementing new technologies in construction projects.

Examinations of emerging technologies such as mobile and wireless are combined with overviews of relevant management theories, and industry case studies, to provide a comprehensive guide suitable for both practitioners and students. Underpinned by research carried out by leading academics in cooperation with practitioners using the latest technologies, this is the most up-to-date and relevant guide to this crucial subject available. The book is essential reading for all practitioners and serious students of management in the built environment.

Dino Bouchlaghem is Professor of Architectural Engineering at Loughborough University, and Director of the Industrial Doctorate Centre for Innovative and Collaborative Construction Engineering, as well as editor-in-chief of the *International Journal of Architectural Engineering and Design Management*. He received the Gold Medal award from the Institution of Civil Engineers for research he published in 2005, as well as a highly commended award from Emerald Literati Network for his research paper 'The Impact of a Design Management Training Initiative on Project Performance' in 2006.

Notes on contributors

Dino Bouchlaghem is Professor of Architectural Engineering at Loughborough University and Director of the Engineering Doctorate Centre for Innovative and Collaborative Construction Engineering. He is the editor-in-chief of the *International Journal of Architectural Engineering and Design Management*. His research interests include collaborative working, sustainable design and construction, and knowledge management.

Patricia Carrillo is Professor of Strategic Management in Construction in the Department of Civil and Building Engineering at Loughborough University. She is Visiting Professor at the University of Calgary, Canada, and the University of Colorado, USA. Her area of expertise is in knowledge management, business performance and IT in construction.

Ashraf El-Hamalawi is a Senior Lecturer (Associate Professor) at Loughborough University. His work has focused on developing computer models capable of providing high-quality practical solutions to civil engineering problems. He is a professionally registered engineer, member of the American Society of Civil Engineers, and chair of the East Midlands branch of the British Institution of Civil Engineers.

Bilge Erdogan is a Lecturer in Construction Management at Heriot-Watt University. She has been working on a number of different research areas including strategic IT management and adoption; visualisation and mobile technologies; collaborative working and collaboration environments; organisational change management, scenario planning and vision development.

Ozan Koseoglu obtained his PhD on Construction Innovation and Technology Management from Loughborough University in the UK. He then joined one of the largest private contractors in UK that also operate in Europe, the Middle East and Australia. He is currently an Assistant Professor in the Civil Engineering Department in the Eastern Mediterranean University and Director of the Construction Management Research Centre (CMRC).

Yasemin Nielsen has over 20 years of research, teaching and industry experience in Turkey, the UK and the UAE. She currently holds a visiting faculty position in Abu Dhabi University. Her research focused on sustainability, strategic management, business and organisational management, visualisation, mobile IT, interoperability, and other construction and project management areas.

Mark Shelbourn is a Senior Lecturer in Construction Management at Nottingham Trent University. He is joint editor of the *Structural Survey Journal*. His research interests include the effect of ICTs on older people; the use of ICTs in the classroom; and collaborative working.

Abdullahi Sheriff is a senior consultant with Happold Consulting specialising in information management strategy for collaboration. Abdullahi's work and research focuses on aligning an organisation's holistic information needs (including enterprise model, content model, technology and change management) to its core business strategy to add strategic value, optimise collaborative working and improve operations, particularly within global multidisciplinary organisations.

Fan Yang is a civil engineer currently working within the Mouchel's Energy Business Unit in the UK. She holds a Master's degree and a PhD from Loughborough University; her doctoral study investigated multidisciplinary collaborative optimisation in building design. Her research interests include innovative design, construction procurement and building information modelling.

Steven Yeomans is the Research Manager in the Centre for Innovative and Collaborative Construction Engineering within the Department of Civil and Building Engineering at Loughborough University. As the former Head of Collaboration for an international engineering organisation and having completed his Doctor of Engineering degree in ICT-enabled collaborative working, he has been an industry advocate of ICT-based collaboration for the past decade.

Collaborative Working in Construction

Edited by Dino Bouchlaghem

Spon Press
an imprint of Taylor & Francis
LONDON AND NEW YORK

First published 2012
by SPON Press
2 Park Square, Milton Park, Abingdon, Oxon OX14 4RN

Simultaneously published in the USA and Canada
by SPON Press
711 Third Avenue, New York, NY 10017

SPON Press is an imprint of the Taylor & Francis Group, an informa business

British Library Cataloguing in Publication Data
A catalogue record for this book is available from the British Library

Library of Congress Cataloging in Publication Data
Collaborative working in construction / edited by Dino Bouchlaghem.
p. cm.
1. Building—Superintendence. 2. Teams in the workplace. 3.
Cooperativeness. I. Bouchlaghem, Dino.
TH438.C6297 2011
690—dc23
2011013341

ISBN: 978-0-415-59699-2 (hbk)
ISBN: 978-0-415-59700-5 (pbk)
ISBN: 978-0-203-84051-1 (ebk)

Typeset in Sabon
by Prepress Projects Ltd, Perth, UK

FSC
www.fsc.org
MIX
Paper from responsible sources
FSC® C004839

Printed and bound in Great Britain by
TJ International Ltd, Padstow, Cornwall

Contents

Figures

Tables

Chapter 1

Introduction

Dino Bouchlaghem and Mark Shelbourn

Background and context

There is a widespread recognition that the construction industry must embrace new ways of working if it is to remain competitive in a global market and meet the needs of its ever demanding clients. Inherent within this drive is a requirement for better collaborative working which is becoming a core part of management paradigms such as concurrent engineering and lean production. Effective collaborative working is essential if design and construction teams are to cover the entire lifecycle of the construction product and take into account not only the primary functions of a built asset but also the production, maintenance and decommissioning processes.

Collaboration in the construction industry has special characteristics in that the formation of the teams and the end product is unique from project to project. Many construction organisations are adopting electronic systems to facilitate collaboration between distributed teams as a means of achieving higher productivity and improving competitiveness through the enhanced quality of their work products (COCONET, 2003). One such approach involves the use of systems that support working in virtual teams that collaborate across geographical, temporal, cultural and organisational boundaries to meet the demands and requirements of the global economy. In this way, project stakeholder interactions can be structured around a virtual work environment with a shared workspace populated with project information. It is, however, sometimes still important to create a feeling of co-presence between non-collocated collaborators if computer-mediated communications are to be successful and broadly adopted. Information-based applications that provide support for the management of individual and shared workspaces are currently widely available.

Despite the big upsurge of interest in partnering and alliances in recent years, there is little knowledge of the nature, feasibility, benefits and limitations of current practices of project stakeholder collaboration (Bresnen and Marshall, 2000). These practices essentially require effective access to information and interactions between project participants and are considered crucial to the competitiveness of large companies (DeRoure *et al.*, 1998). Communication is an important part of collaboration in construction; an efficient and systematic exchange of information between designers, contractors, consultants, and sub-contractors is considered essential if construction projects are to be completed on time and within budget (Cohen, 2000; Atkin *et al.*, 2003; Emmitt and Gorse, 2003). Design teams, in particular, increasingly need to make significant efforts to establish and maintain efficient communication and good coordination of information to accommodate the complex multidisciplinary collaborative design environment. Software tools and hardware solutions that support such distributed design teams have therefore become a necessity rather than just a trend (Peña-Mora *et al.*, 2000). The effective management of design information and design changes is now regarded as an essential element for the success of construction projects. Collaborative design is discussed in detail in Chapter 6, which also introduces a computer-based method that supports the multidisciplinary collaborative design development.

The construction process, on the other hand, often suffers from the lack of collaboration and inefficient communication between design offices and construction sites. The most common cause of construction site problems is acknowledged to be the poor coordination of design information. Furthermore, the main causes for project delays from the contractors' point of view are usually reported to be delays in receiving design information, long approval timescales, design mistakes, conflicts between design documents, and change orders (Kumaraswamy and Chan, 1998). Communication and collaboration between designers and construction teams still mostly rely on face-to-face meetings, and basic media such as telephone, email and fax transmissions. Site-based engineers usually use a number of drawings to perform a single activity. Therefore, fast access to design information through digital media on construction sites can significantly improve the efficiency of information exchange between the job site and the design office and hence facilitate site operations. Moreover, design details are not always simple and easy to interpret, and engineers can spend considerable time trying to understanding complex details; this can sometimes result in the actual construction being different from the design specifications. These issues are discussed in Chapter 5 of this book together with the results of an investigation into the potential that mobile technologies can offer to improve collaboration between designers and construction teams on site.

Collaboration and information technology

Much of the recent development on collaborative working has focused on the delivery of technological solutions, concentrating on the web (extranets), computer-aided design (visualisation) and information/knowledge management technologies (Kvan, 2000; Woo *et al.*, 2001; Bouchlaghem *et al.*, 2005; Tan *et al.*, 2010). The collaboration merits of information and communication technology (ICT) tools are usually associated with their capabilities in supporting a high level of interaction, many-to-many communication and information sharing, in a group of known users, across the hierarchical, divisional, time and geographical boundaries (Karsten, n.d.). However, it is now recognised that good collaboration does not result only from the implementation of information systems, as such focused approaches are proving to be less than successful (Vakola and Wilson, 2002; Ferneley *et al.*, 2003). On the other hand, approaches that exclusively focus on organisational and cultural issues would not reap the benefits gained from the use of ICT, especially in the context of distributed teams that are the norm in construction.

The business and cultural environments within which collaboration takes place are important factors when introducing collaboration technologies. However, as with other efforts that involve the implementation of ICT in projects and businesses, there has been a traditional focus on the technological aspects (Kvan, 2000; Faniran *et al.*, 2001; Woo *et al.*, 2001) with limited consideration of the effects that these implementations have on the users and the organisations they work for (Alvarez, 2001; Vakola and Wilson, 2002; Ferneley *et al.*, 2003). Managing collaborative working in organisations cannot be achieved without the use of ICT; equally, technology implementations that concentrate solely on the social and economic aspects in organisations will also encounter difficulties (Koschmann *et al.*, 1996; Loosemore, 1998; Winograd, 1998; Eseryel *et al.*, 2002).

Introducing ICT into engineering organisations affects many cultural and behavioural aspects of the work environment (Credé, 1997; Proctor and Brown, 1997; Cheng *et al.*, 2001). There are cultural differences around virtually every corner in any typical organisation (Haque and Pawar, 2003). Taking these differences into consideration is important when managing the change required for a smooth transition into any new working practices resulting from the implementation of new systems. Individuals can be apprehensive when confronted with technological change (Manthou *et al.*, 2004). The need for change combined with the fear of the unknown when faced with new technologies can be unsettling for many professionals. Experience shows that technology does not always lead to improved practices at first, may not work as intended, and can be seen to have negative impacts on some individuals within organisations (Bartoli and Hermel, 2004). These negative results are particularly common when introducing new and advanced technologies that necessitate significant changes to accommodate their operational needs.

Disconnections can exist between business departments, managers and, most critically, ICT users and ICT developers. In order to introduce information systems change into an organisation effectively, it is necessary, as in any action that requires business change, to carefully consider the end users' views (Finne, 2003). These users, especially operational managers, function and think using situational orientation as they react to the changing stimuli and contingencies around them. Technicians, on the other hand, are linear and rely on technical capabilities to solve problems. Understanding the differences between the two modes of thought and orientation is the first step towards bridging the gap that naturally exists between managers (system users) and technicians (developers) (Zolin *et al.*, 2004).

New technologies can trigger different reactions from individuals as a result of the diverse and subjective interpretations and perceptions of the level of change required; this does not always result in resistance or rejection. Managing change when introducing computer technologies should therefore involve balancing technological, organisational and human factors (Korac-Kakabadse and Kouzmin, 1999; Chesterman, 2001; Smith, 2003). In addition to individuals' and groups' reactions, managers also have to take into consideration the relationships between information technology, information management and organisational change (Patel and Irani, 1999; Claver *et al.*, 2001). These aspects of information and change management are particularly important in the construction sector and will be covered in detail in Chapters 2 and 7 of this book. In addition, a framework for the management of change in organisations and a structured approach for the design, selection, implementation and maintenance of a holistic information strategy are presented in Chapters 4 and 8 respectively.

An integrated approach to collaborative working

There has been a rapid growth in the development of collaboration tools and systems in recent years, especially in the areas of communication, visualisation, and information and knowledge management; however, the uptake and implementation of these tools has been rather slow and with mixed levels of success. In the absence of well-defined strategies that take into account the organisational, project and users' requirements, choosing and implementing collaborative systems within the construction industry is sometimes done in an ad hoc manner. It is believed that a well-defined and mapped methodology for collaborative working will maximise the use of and benefits from ICT-based collaboration systems.

Evidence shows that the current collaboration tools landscape is improving but at the same time still fragmented and lacking integrated solutions (COCONET, 2003). This may be because the scale and scope of cooperative tasks is increasing, and as a consequence the use of collaborative systems is becoming more pervasive (Dustdar and Gall, 2003).

This book imparts an integrated view on collaborative working in construction, and addresses the technological, business and process factors associated with effective collaboration, together with the change management required to facilitate the introduction of collaboration systems. An integrated framework, which forms the core of this book, is introduced in which organisational priorities for collaborative working are considered together with project needs, users' requirements and technologies in a decision-making framework that aims to support organisations in the strategic planning and implementation of effective collaborative working policies and practices. When carefully planned and if based on informed decisions, it is believed that these policies and practices will help organisations improve their collaborative working, achieve full benefits from it and maximise the use of tools and techniques available. Particular attention is given to the challenging requirements of distributed, heterogeneous and transient construction project teams together with the need to support the planning for all aspects of collaboration between team members and across all stages in the project delivery process. The Planning and Implementation of Effective Collaboration in Construction (PIECC) framework is presented in Chapter 3.

Acknowledgements

The research on which this book is based was funded by the Engineering and Physical Sciences Research Council (EPSRC). Special acknowledgements go to those who contributed and gave support to various aspects of the work and in particular the group of professionals from the construction industry who formed part of the core team that developed the PIECC framework and handbook. These include: Alan Blunden, Rennie Chadwick, Clive Cooke, Richard McWilliams, Gary Rowland and Paul Waskett.

Chapter 2

Collaboration

Key concepts

Mark Shelbourn, Abdullahi Sheriff, Dino Bouchlaghem, Ashraf El-Hamalawi and Steven Yeomans

Introduction

Collaboration is the collective work of individuals and groups undertaken with a sense of common purpose and direction within a shared environment that combines physical, digital and virtual resources. In collaborative work relationships, awareness of and familiarity with the environment develop through interaction with other stakeholders and access to information and knowledge. The collaboration process is characterised by the way it affects individuals and groups who manage their work so that it meets the collective needs through the coordination of their efforts to create positive changes and achieve common goals that become primary forces for generating commitment. The following standard definition of collaboration will be used in this book:

> Collaboration is an activity in which a shared task is achievable only when the collective resources of a team are assembled. Contributions to the work are coordinated through communications and the sharing of information and knowledge.

This chapter discusses the key concepts and types of collaborative working with a focus on what forms effective collaboration and the barriers in the way of achieving it. The characteristics of collaboration in construction are reviewed together with the tools, techniques, resources and processes that support its planning and implementation within project environments.

Collaboration in construction

It is widely accepted that improving productivity and performance in the construction industry requires an integrated collaborative approach to project delivery, and managing information effectively is crucial to achieving this. The multidisciplinary nature of the construction industry with its often bespoke and transient projects makes the nature of and need for collaborative working different from other fields. Construction projects rely on collaborative working between a wide range of disparate professionals working together for a relatively short period to design and deliver a project; much of this process is based on a traditional sequential approach in which many of the participants often work independently, make decisions that inevitably affect others and then come together in face-to-face meetings. Furthermore, various individual stakeholders collaborating on a project also have differing organisational objectives; rarely have a common project aim; and have also received diverse educational upbringings that use different terminologies (unlike a group of medics collaborating to aid a patient). Although this ad hoc approach has entrenched the practice of collaborative working, it has also reinforced traditional disciplines to the extent that, on many projects, an adversarial environment prevails and the fundamental ethos of collaboration is not fully evident (Anumba *et al.*, 2000). The situation is also further exacerbated by the use of collaborative practices alongside traditional approaches and adversarial contractual arrangements. Professional practices that recognise the need for collaboration generally adopt one of the following two methods to facilitate it (Kalay, 1999):

1 hierarchically partitioned – based on a contractual form, in which one team member (often the architect) takes leadership of the group (consultants and sub-contractors); considered an efficient process but less than optimal for collaboration;
2 temporally partitioned – in which responsibilities follow the typical 'over the wall' practice, shifting from one professional group to another; again considered detrimental because of the implication of ownership and quality.

Based on the above, it is feasible to argue that, although the fragmented nature of the industry makes collaborative ventures more important, it also makes it harder to realise. Simply bringing a group of people together does not necessarily ensure that they will function effectively as a team. Lack of organisation, misunderstanding, poor communication and inadequate participation can all lead to problems and difficulties. Collaboration in construction is therefore an extremely complex process that requires a high level of strategic planning.

The sequential or 'over the wall' approach to project development has resulted in numerous problems for the construction industry (Anumba *et al.*, 1997), including poor integration, coordination, communications and collaboration leading to fragmentation and ultimately poor value for the client.

Collaboration, cooperation and coordination

The terms 'collaboration', 'cooperation' and 'coordination' are often used interchangeably; there are, however, some fundamental characteristics that differentiate them:

- *Cooperation* is characterised by informal relationships that exist without a commonly defined mission, structure or dedicated effort. Information is shared as needed and authority is retained by each organisation so there is virtually no risk. Resources are separate, as are rewards.
- *Coordination* is characterised by more formal relationships and understanding of compatible missions. Some planning and division of roles are required, and communication channels are established. Authority still rests with the individual organisation, but there is some increased risk to all participants. Resources are available to participants and rewards are mutually acknowledged.
- *Collaboration* connotes a more durable and pervasive relationship. It represents 'full commitment to a common mission. Authority is determined by the collaborative structure. Risk is much greater' (Mattessich and Monsey, 1992: 39).

Collaboration requires a greater commitment to a common goal than cooperation. It also involves a collective sharing of risks and therefore requires a higher level of trust between stakeholders.

Types of collaboration

There are four different modes of collaboration, which relate to the type of interaction and pattern of communication between participants in a project (Figure 2.1; Anumba *et al.*, 2002):

1 *Face-to-face* collaboration normally involves physical meetings in a shared venue such as a meeting room where participants interact in real time.

	Same Time	Different Time
Same Place	**Face-to-Face Collaboration** *e.g. in meeting rooms where participants engage in face-to-face discussions*	**Asynchronous Collaboration** *e.g. where communication is conducted via some form of notice / bulletin board*
Different Place	**Synchronous Distributed Collaboration** *e.g real-time communication using any of a vast array of current ICTs i.e. telephone, video conferencing, electronic group discussion etc.*	**Asynchronous Distributed Collaboration** *e.g communication via the post such as letters, fax machines, telephone messages / voice mail, pagers, email etc.*

Figure 2.1 Modes of collaboration.

2 *Asynchronous* collaboration is conducted in a shared location where participants do not interact at the same time; it requires electronic media such as notice/bulletin boards.
3 *Synchronous distributed* collaboration involves real-time interaction in which participants are not collocated but use communication technologies such as computer-mediated meetings, videoconferencing and electronic discussion groups.
4 *Asynchronous distributed* collaboration is when participants interact from dispersed locations but not in real time, using various communication media such electronic mail.

Collaboration can also be classed as:

- *mutual*, in which the participants have the same responsibility and level of input to a work task;
- *exclusive*, in which the participants have exclusive responsibility for different parts of the work schedule but occasionally interact to coordinate the different tasks; and
- *single point control*, in which the participants delegate the responsibility to someone to lead the process and allocate responsibility.

Regardless of the mode or approach, it is argued that effective collaboration necessitates successful and efficient sharing of knowledge, negotiation, coordination and management of activities (Lang *et al.*, 2002). It should also be founded on good communication, as it is the means by which the intents, goals and actions of each of the participants are made known to others (Kalay, 2001). Consequently, organisations need to formulate a well-defined communication strategy to ensure delivery of the appropriate information, to the correct recipient and in real time, to facilitate truly effective collaborative working. In summary, successful collaboration requires effectiveness in a number of areas:

- coordination;
- negotiation;
- communication of data, information and knowledge;
- agreeing shared vision and goals;
- planning and management of activities and tasks;
- adopting common methods and procedures;

and can deliver the following benefits:

- added value to a project;
- increased revenues and profits;
- improved business efficiency;

- improved productivity of individuals as a result of being part of a team;
- improved customer satisfaction;
- enhanced collective image of the groups within the collaboration partnership.

Challenges for collaboration

Collaboration involves multidimensional and cross-disciplinary tasks that require interactions between stakeholders and the exchange of information across organisational boundaries using shared information and communication technology (ICT)-based collaborative environments. Owing to the complexity associated with meeting the above requirements, some of the possible difficulties commonly experienced include (COCONET, 2003):

- time and data losses during information exchange;
- incompatibility between communication infrastructures used by different participants;
- misunderstandings caused by ill-defined information;
- complex iterative negotiations when solutions conflict;
- lack of effective and efficient tools for organising and exchanging project information;
- the coordination of complex work processes requiring input from different organisations;
- lack of awareness of the characteristics of others within the social, ethical and technological contexts;
- lack of consistency in collaboration support for 'collaboration spaces' and multimedia conferencing;
- lack of support for creativity, discourse management and conflict resolution.

Research on stakeholder collaboration in construction is often insufficiently informed by the many social science concepts and theories (relating to motivation, teambuilding, organisational culture and the like) that are central to an understanding of cooperation and trust between organisations. Moreover, with many small and medium-sized enterprises (SMEs) involved in construction projects often not actively included in agreements to collaborate on projects, they perceive their involvement in collaboration as a complex and costly process that brings little benefit to their own businesses. Therefore, if collaboration is to work, contractors need established partnering arrangements with suppliers whom they work with regularly.

If collaboration is to become successful in construction then the drive for change must come from high in the supply chain (i.e. from the client or contractor) and be followed right down to the suppliers. All collaboration partners need to be empowered to freely interact with each other, find and use resources as and when needed, and bring in experts in required domains.

Types of collaborative organisations

Collaboration should occur naturally but organisations tend to create barriers keeping this from happening. For example, in traditional functional organisations, often a decision has to be escalated from worker to supervisor to manager in one function, then across to a manager in another function and down to a supervisor back to worker before a final decision is made and communicated. The result is a loss of decision-making quality and time. Knocking down functional barriers and allowing workers to talk directly to relevant parties and make their own decisions (when possible) enhances natural collaborative processes and results in better and faster decisions.

Figure 2.2 illustrates organisation types as a function of formal and informal collaboration practices. Formal forms include temporary or permanent teams, single-function or multifunction teams, collocated or distributed teams, and cross-functional or function-specific teams. Informal forms include communities of practice, learning communities and the 'water cooler'. Both formal and informal forms depend on structural support and cultural changes, but perhaps to different extents. Ideally, an organisation promotes both formal and informal forms, becoming what can be called a collaborative organisation. Each of the organisation types in Figure 2.2 is described briefly in Table 2.1.

Effective collaboration

Effective collaboration can enhance organisational learning by encouraging individuals to exercise, verify, solidify and improve their knowledge

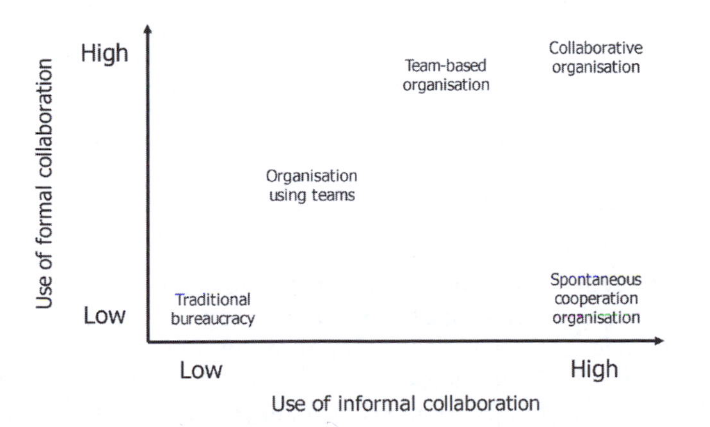

Figure 2.2 Organisation types as a function of use of formal and informal collaboration.

Table 2.1 Organisation types and their ability to collaborate

Type	*Description*
Traditional bureaucracy	No teams at any level Norms, rules, and procedures inhibit informal collaboration (e.g. discussing problems over the water cooler is seen as wasting time and is punishable by the rules) Focus of systems (e.g. rewards and compensation, performance management) is on the individual Individuals are usually organised in functions (e.g. engineering, production) High level of hierarchy in reporting structure
Organisation using teams	Some teams used at any level Norms, rules, and procedures inhibit informal collaboration Focus of systems is on the individual Individuals are usually organised in functions Medium to high level of hierarchy in reporting structure
Spontaneous cooperation organisation	Few to no teams used at any level Norms, rules and procedures support informal collaboration (e.g. a norm that individuals consult with each other when they need help) Focus of systems is on the individual Individuals are usually organised in functions Medium to low level of hierarchy in reporting structure
Team-based organisation	A variety of team types are used as the basic units of accountability and work; workers and managers are organised in teams Norms, rules and procedures do not actively support informal collaboration Focus of systems is on individual, team and organisation Teams are usually organised around processes, products, services or customers Low level of hierarchy in reporting structure
Collaborative organisation	A variety of team types are used as the basic units of accountability and work; workers are organised in teams; managers may or may not be organised in teams Norms, rules and procedures actively support informal collaboration (e.g. common spaces such as lounges are created and employees are encouraged to meet there to discuss issues) Focus of systems is on individual, team and organisation Teams and individuals are usually organised around processes, products, services or customers Low level of hierarchy in reporting structure

through interaction and information sharing during the collective problem-solving and decision-making process. This is often supported with tools and technologies that facilitate communication and knowledge creation, storing and dissemination. Hence, in order to achieve effective collaboration, organisations must harmonise, bring together and carefully align the three key strategic areas of business, people and technologies (Figure 2.3).

Collaboration enables participants to build capacity in order to complete

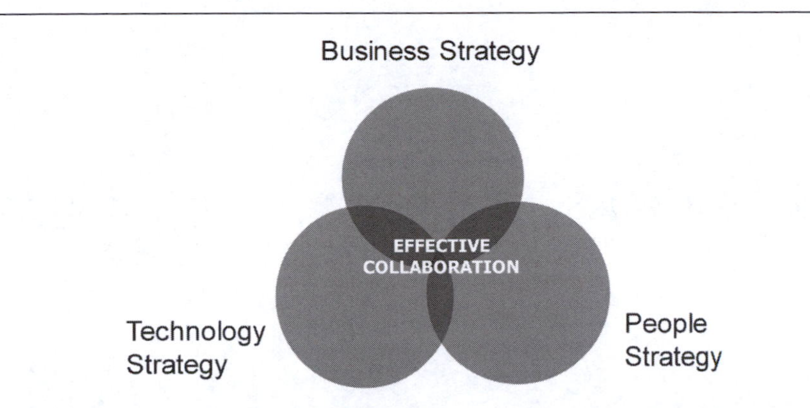

Figure 2.3 Key areas for effective collaboration.

a set of tasks that one sole organisation would find difficult to cope with. Through a shared vision, it can prevent fragmentation, duplication and distrust. One of the main objectives of collaboration is the efficient use of the combined resources of all stakeholders in the supply chain while sharing project risk factors across multiple domains and enhancing staff and organisational motivation.

Good collaboration requires trust between participants, clear processes and efficient communication infrastructures supported with appropriate technologies. To be successful, a collaborative project must establish a definition of the team, identify their outcomes, ensure there is a purpose for the collaboration and clarify interdependencies between members. There are six key factors that are deemed critical for effective collaboration in construction (Figure 2.4); these are linked to the three broad strategic areas above:

1. *vision* – all members of the collaboration should agree on its scope, aims and objectives;
2. *stakeholder engagement* – collaboration leaders need to ensure that all key participants are consulted on the practices to be employed during the collaboration;
3. *trust* – time and resources are needed to enable all participants to build trusting relationships;
4. *communication* – a common means of communication should be decided by all key participants in the collaboration;
5. *processes* – how the collaboration is to work on a day-to-day basis, in relation to both business and project, should be known by all key participants;
6. *technologies* – an agreement on those technologies to be used is required to ensure the collaboration is easily implemented and managed.

Perhaps the most important overarching aspect of effective collaboration is that working collaboratively often means new ways of working for

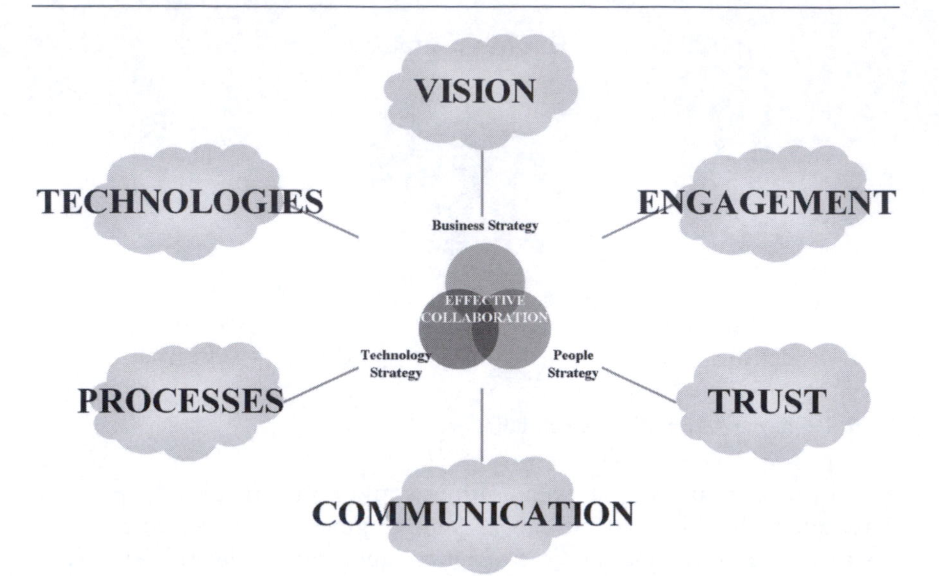

Figure 2.4 Areas to be addressed to enable effective collaboration.

many or all of the participants involved. Effective collaboration is achievable through the innovative design and development of a balanced strategy that does not rely solely on sophisticated information and communication technologies. As yet there is little support available that prescribes to managers effective ways of implementing and managing collaborative project environments. This lack of an integrated approach for collaborative working has been identified and addressed by the Planning and Implementation of Effective Collaboration in Construction (PIECC) project presented in Chapter 3, which aimed to develop a framework that takes advantage of the benefits of a more targeted use of ICTs which are better aligned to the people and business strategies within organisations.

Barriers to effective collaboration

It is widely accepted that people and relationships are considered to be at the heart of collaboration success and that the lack of continuity of relationships (at company, team and individual levels) frequently undermines achieving the full benefits of collaboration and transferring experience across projects. The strategic planning of longer-term, mutually beneficial customer–supplier relationships is needed to realise the benefits of collaboration in concurrent engineering (May and Carter, 2001). These relationships are difficult to achieve and even more difficult to maintain and sustain.

Although there is a potentially important symbiotic relationship between internal and external collaboration processes, difficulties can occur if project team cultures clash with the wider organisational values and norms (Bresnen and Marshall, 2000). Getting the people to use the tools that make collaboration possible is another major barrier to effective collaboration. This is often caused by a combination of human and management factors including resistance to change due to lack of training and change management support, and the lack of adequate management and support for the collaboration tools and technologies.

Some of the other common barriers to effective collaboration can be summarised as:

- collaborating organisations having different vision, mission, goals and priorities;
- organisational 'culture' and methods of communication are often different;
- a lack of focus and consensus on the delegation of tasks;
- an imbalance of resources – time, money, human (frequent turnover of participants) etc.;
- confidentiality, intellectual property and legal considerations;
- technological incompatibility;
- a lack of understanding of the expertise, knowledge and language of the other collaborating participants.

Information and communication technologies

Information and communication technology advances have rapidly changed the ways firms manage and operate their businesses. ICT enables firms to better manage their business processes through new and improved business models, and to strengthen their outreach activities and interaction with the supply chain. Information-intensive business organisations are utilising ICT to create and manage new knowledge, exchange information, and facilitate inter- and intra-organisational collaboration (Lee, 2004).

It has become accepted practice to use the term 'collaborative environments' to denominate the computer systems which support communication and exchange of information between project stakeholders within the collaboration activities. Hence, the implementation of ICT can improve collaboration in an organisation if the following conditions are present:

- Organisation members need to collaborate.
- Users understand the technology and how it can support collaboration.
- The organisation provides appropriate support for the adoption, implementation, and continued use of the technology.
- The organisational culture supports collaboration.

Collaboration environments provide tools that support the sharing of information between different applications, concurrent access to data and the exchange of information objects through a network, in order to enable collaboration between different users and support the so-called virtual teams (Bouchlaghem *et al.*, 2006).

Collaboration tools include intranets, extranets, enterprise information portals, knowledge management applications and others. These tools should complement but not replace traditional methods of communication; their application is about reducing administrative burden and improving communication speed.

Online collaboration through internet-based conferencing systems can reduce the need for physical project meetings, leading to significant savings in time and resources. Development engineers from the automotive industry estimated that, on average, about 70 per cent of their time can be spent on activities involving external suppliers. By using online synchronous collaboration tools, many design and engineering issues can be resolved as and when they arise, resulting in a more responsive mode of working. While online collaboration enables shared understanding of the overall design issues with input from multiple engineers and designers, offline activities can focus on the development of component parts taking into consideration the collaborative feedback.

Technologies can be chosen and customised to support specific areas of collaboration. For example, different types of technologies can support each of the four types of collaboration described above, as illustrated in Figure 2.5.

The key to successful technology-supported collaboration does not depend solely on the technology implementation and use. It also depends on an organisation's ability to adopt new ways of working. The best technology available will not inherently enhance people's ability to work together. Neither will it improve individual and team productivity, unless the people using it have made the necessary behavioural changes. There is evidence to suggest that inadequate or inefficient teamwork can be made worse by technology adoption. One of the greatest challenges facing individuals joining a geographically distributed team is isolation. For some people it represents freedom and autonomy as they learn to adapt and flourish in this environment, whereas others find it difficult to adjust. Reliance on face-to-face methods of collaboration means that when it is taken away people can suffer a crisis of work identity (Conner and Finnemore, 2003). Table 2.2 summarises the barriers, drivers, requirements and benefits of the use of ICT in collaborative working.

Information management and collaboration

The design and construction process brings together numerous stakeholders and participants often from multiple organisations working together as a 'temporary enterprise' through a procurement process to develop and implement solutions to meet client needs. The solutions resulting from

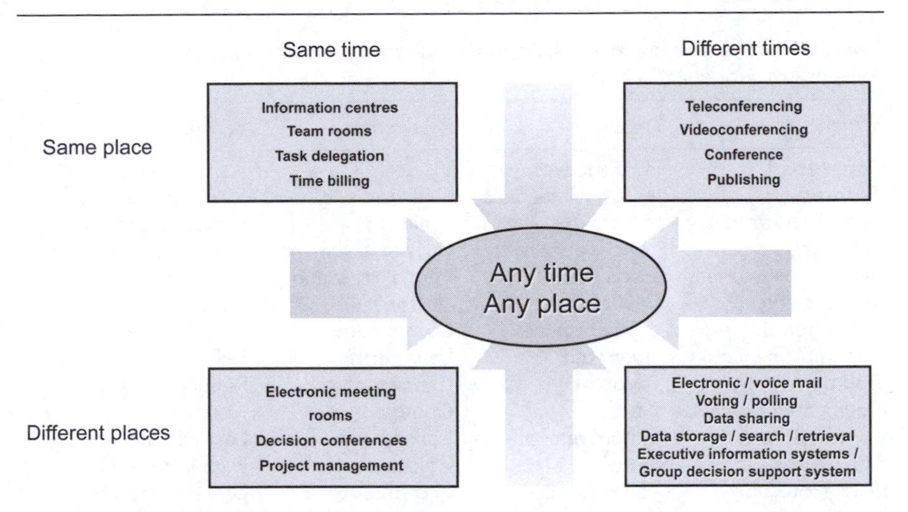

Figure 2.5 Possible technologies to be used for the four types of collaboration.

this often complex interaction emerge through the creation and continuous exchange of sometimes large volumes of information, which is usually defined as the product of the contextual understanding and interpretation of data. It is the essential medium through which knowledge, expertise, judgement, emotions and decisions held by individuals are expressed, shared and communicated with others (Davenport and Marchand, 2000). Information management (IM) encapsulates all the activities that support the information lifecycle from creation, representation and maintenance through to reuse (Hicks *et al.*, 2006). An information intelligent organisation is one which understands the value of information and can successfully search, find, assemble, analyse, use and reuse all forms of information products required for any of its tasks (Evgeniou and Cartwright, 2005). This is particularly important as competitive advantage today makes information central to doing business and obtaining operational efficiency (Christian, 2002; Hicks *et al.*, 2002, 2006; Chaffey and Wood, 2004; Laudon and Laudon, 2009). Being information intelligent requires a more strategic view of information as a corporate asset, aligning the information needs of the organisation to its business processes (Buchanan and Gibb, 1998). It requires a fundamental rethink of information, its position within the organisation and its potency as a means of securing long-term competitive advantage. It also requires information to be viewed in a holistic manner, balancing an appreciation of technologies with the capabilities of people within the business to harness and use the information to improve performance (Marchand, 2000). To clarify its scope, it is necessary to differentiate between IM and associated concepts in both research and practice.

Recent research on IM has focused on the project environment and highlighted the need for improved collaboration and coordination between stakeholders (Caldas and Soibelman, 2003; Peansupap and Walker, 2005).

Table 2.2 Summary of the barriers, drivers, requirements and benefits of ICT-enabled collaborative working

Barriers	*Drivers*	*Requirements*	*Benefits*
Commercial constraints	EU directives	Vision	Attainment of industry targets
Lack of strategic planning	Industry initiatives	Collaborative mindset	Increased profitability
Lack of commitment time/energy/resource	Industry improvement	Early involvement	Efficient collaboration
Late implementation	Sustainability	Mutual trust and respect	Improved communications
Traditional contacts	Client requirements	Communication	Integrated supply chain
Fragmented process, professions and disciplines	Integrated approach	Long-term relationships	Reduced and shared risk
Data protection	Process improvement	Common processes	Less conflict
Intellectual property rights	Better collaboration	Collaborative contractual arrangements	Higher quality
Legal admissibility	Integrated systems	Transparency	Lower costs
Fear of change	Innovation and new technology	Open negotiation	Fewer errors, less rework and waste
Resistance to change	Globalisation	Adaptability	Increased efficiency
Fear of failure	Knowledge management	Change management	Greater predictability
Different cultures		Standardisation	Improved decision taking
Different languages		Best value not lowest price	Reduced programme time
Time zones		Performance measurement	Shared understanding
Diverse company values		Continuous improvement	Consensus
Diverse company procedures		ICTs	Design certainty
Lack of understanding		Education	Integrated working
Lack of experience		Training	Process improvement
Lack of education		Legal clarification	Change management
Lack of training			Resource management
Lack of ability			Improved work environment
Human behaviour			Rapid information exchange
Lack of trust			Better data management
Multiple standards			Redesign on PC not site
Poor standards			Remote and mobile working
Poor interoperability			Overcoming geographical barriers
Information overload			Improved health and safety
Vendor commercial interests			Whole lifecycle analysis
Lack of investment			
Poor adoption rates			
Inaccessible information			
Skill shortage			
Rapid ICT product change			

As little research has been conducted on IM from an organisational perspective there is limited clarity on its specific nature, needs, drivers and barriers within organisations. Improving collaborative working is one such driver but remains a broad field with diverse yet interrelated and often interdependent themes, only one of which is related to IM (Jorgensen and Emmitt, 2009). Although construction organisations' primary focus is on projects, a merely project-centric view does not represent all the information created, shared and managed within organisations nor does it empower organisations to manage cross-project information. Managers also lack an understanding of the broader issues of IM, the type of information various people within the organisations need and want, and how to effectively implement such a strategy to support their organisations.

Whereas an effective collaboration strategy includes the process, people and technological dimensions shaped by a clear global vision, little emphasis is usually placed on 'information' or 'content' and its role in defining the purpose for collaboration and indeed an effective collaboration strategy within organisations. The management of information is instead largely viewed as synonymous with technology and/or systems. However, the growing realisation of the criticality of leveraging information to improve operational effectiveness and collaborative working within organisations demonstrates a need for organisations to develop effective IM strategies to underpin business strategies. This section describes the concept of a holistic approach to IM as an essential element in developing and implementing effective collaborative working in organisations.

Information systems and information technology

Information management (IM), information systems (IS) and information technology (IT) are sometimes referred to almost interchangeably, which shows an apparent lack of understanding of the wider context of IM. Efforts to integrate IM with organisational processes often focus on IS and IT solutions without an overall integrated IM strategy. The result is largely technologically biased and excludes the full organisational dimension. Although all three areas focus on information (as shown in Figure 2.6), the emphasis of each is different, making them distinct fields with different requirements. The focus of the three information streams can be summarised as:

1 *IT* – primarily concerned with the infrastructure needed to manage information, ranging from desktop-based infrastructure to servers and networks, with emphasis placed on reliability, responsiveness, flexibility and ease of use of the various technologies;
2 *IS* – focused on the software applications which perform defined business functions ranging from design, manufacturing and production to

accounting, human resource management and other associated processes within the organisation; and

3 *IM* – concerned with content and the bits of information required to carry out distinct tasks/processes. It is strategy and process driven, aligned with the various business units across the organisations. The emphasis here is on managing and leveraging content to support business processes.

Each stream, with its distinct paradigm, enables organisations to gain competitive advantage through technology (IT), software (IS) or information (IM). However, interrelationships and interdependencies do exist between them and it is often impossible to consider one of them in isolation. A successful IM strategy must often be accompanied by successful IT and IS solutions as they form the core media through which information is created, shared and stored. However, a focus on IT or indeed IS does not imply a focus on IM as neither IT nor IS focuses on the content or information

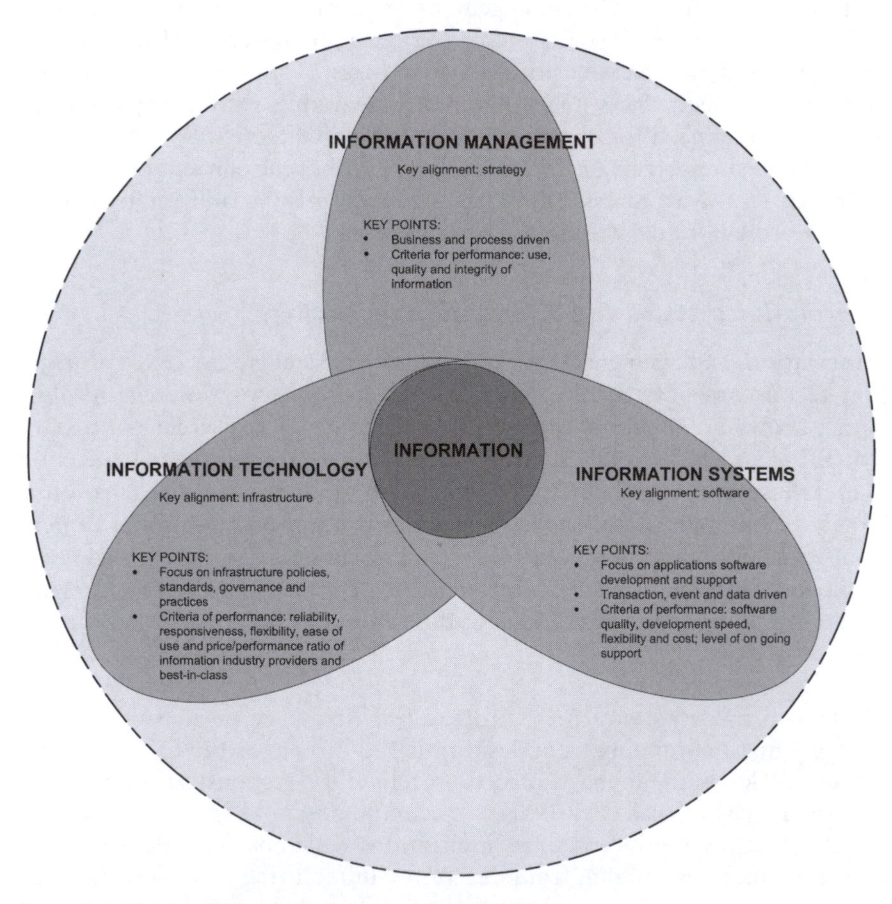

Figure 2.6 IM, IS and IT alignment (from Marchand, 2000).

which an organisation creates or uses; its alignment with the organisation's strategy; or the behavioural dimensions of managing information (all of which are the focus of IM).

An integrated approach to information management

A holistic approach to IM involves the combination of strategies, tools, processes and skills that an organisation needs to manage all forms of recorded information through its complete lifecycle supported by necessary technological and administrative infrastructures. Such an approach requires an appreciation of what information means to an organisation, and how the organisation can best use, structure and exploit it to achieve the desired results. Numerous technologies do exist which aim to enable this; however, critical to the success of a holistic approach is the emphasis on corporate-wide strategies and policies which guide the use and implementation of the appropriate technology. It is often reported that focusing solely on systems and technologies has little impact on productivity unless accompanied by operational changes in processes irrespective of a company's size, location, sector or past performance. Thus a holistic approach, in which information needs are aligned with business processes and technology, would have more impact and adds greater value. Like any successful strategy, this should remain dynamic and adaptable enough to support evolving business goals. A number of drivers were identified in the literature for a holistic approach to managing information. These include (Moore and Markham, 2002; Sprehe, 2005):

- the need to improve collaborative working and knowledge sharing across organisations and with clients;
- the realisation that, in a knowledge-based economy, information is a key corporate asset and as such is crucial to improving competitiveness;
- the need to improve consistency across the organisation and reduce waste (in time and resources) associated with the duplication of information;
- the need to ensure regulatory compliance;
- the need to improve productivity and manage business risks;
- the need to enable greater innovation and value creation across the organisation.

Components of a holistic approach

A holistic approach to IM consists of four key components, all essential and equally important to ensure the approach is contextual, appropriate and implemented effectively (Marchand, 2000; Paivarinta and Munkvold, 2005; Bridges, 2007). These, as illustrated in Figure 2.7, are:

- content model;
- enterprise model;
- systems and technology; and
- implementation and change management.

Content model

The content model refers to the nature of the content including its structure, attributes and fit within the organisation's needs (Grosniklaus and Norrie, 2002; Paivarinta and Munkvold, 2005; Rockley, 2006). Simplistically put, 'content' is a 'bit' of information created by an author (e.g. text, drawings, charts, graphic images) contained within information products (Rockley *et al.*, 2003). Therefore, a document can be described as an information product with specific content. The focus of IM is not on managing information products in the form of electronic files or paper per se, but on the content contained within each product. Therefore the key to defining a suitable content model is deciding on an appropriate level of granularity for the organisation's needs. Granularity refers to the lowest level where a piece of content can be divided while still remaining meaningful and manageable (Rockley *et al.*, 2003). Differences in content granularity have given rise to two distinct approaches to content management in both research and practice, each with its benefits and limitations. These are:

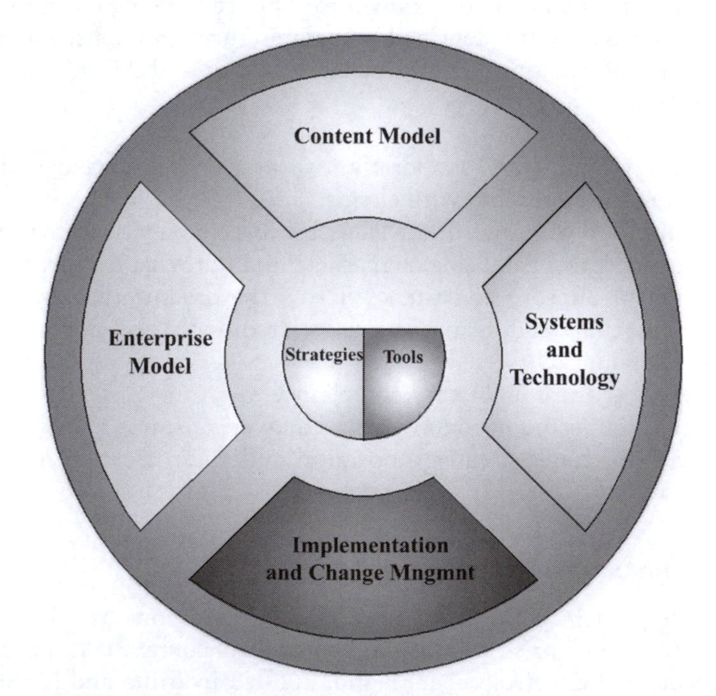

Figure 2.7 The four components of a holistic approach to information management.

1 *The integrated document management approach*: Here, information products such as documents are treated as individual autonomous entities and the focus is to support the information product through its lifecycle (Bjork *et al.*, 1993; Stouffs *et al.*, 2002). This includes systems such as document management (DM), electronic document management systems (EDMS) and extranets.
2 *The model-based approach*: Here, information is created and maintained as granular pieces of content (and not documents). Information products such as documents or drawings are then assembled from this dynamic content in the most appropriate format when required (Rockley *et al.*, 2003; Hamer, 2006). Such concepts include building information modelling (BIM) and content management (CM).

Individual organisations' requirements and needs would dictate which approach is most suitable. Information audits are conducted to gain a clear understanding of the exact nature of an organisation's information needs in line with its business strategy. Two other critical aspects of a content model are the metadata and taxonomies.

Metadata are additional pieces of data or attributes which describe the context, content and structure of a piece of data, content, document or other bit of information and their management through time (International Standards Office, 2001). The use of metadata within a content/document library allows information to be more easily found, its source determined and its context understood, easing interpretation and facilitating reuse (Bentley, 2001; NISO, 2004; Day, 2006; Paganelli *et al.*, 2006). Whereas resource discovery remains one of the principal functions of metadata, others may include provenance, technical specification, functionality, terms of use, administration, and the indication of linkages or relationships (Technical Advisory Service for Images, 2006). Current standards are predominantly bibliographic in focus with very limited research on the use of metadata within organisations (Karjalainen *et al.*, 2000; Murphy, 2001; Paivarinta *et al.*, 2002). A metadata standard is required to underpin a holistic approach to managing information. It creates a unifying framework for integrating multiple systems and improving the structuring, sharing and retrieving of information across the organisation. International standards for descriptive metadata exist, the most prominent of which are ISO 15836 and BS1192:2007 (specific to the construction industry).

Taxonomy is the logical conceptual structuring of information within a given environment (Woods, 2004; EEDO knowledgeware, 2006). It provides the basis for users to access and navigate through the content collections, allowing organisations to make seemingly significant volumes of content readily accessible (Munkvold *et al.*, 2003; Hienrich *et al.*, 2005; Paivarinta and Munkvold, 2005). There is a strong relationship between taxonomy and metadata, with a taxonomy typically built with appropriate metadata (Paivarinta and Munkvold, 2005). Metadata are simply descriptive information having no associated hierarchy or relationships. Taxonomies leverage metadata to organise content and create associations between attributes, easing search and retrieval of vital information (Woods, 2004; Hienrich *et al.*, 2005). Similar to metadata, designing a suitable taxonomy requires a holistic vision of the content to be managed, potential users (internal and external) and the overall business processes (Gottlieb, 2005). The term 'information architecture' is also used to describe taxonomy and the structuring of content. It refers to the set of principles guiding the organisation, navigation and searching of information to enable its management and retrieval (Dong and Agogino, 2001; Dilson, 2002).

Although taxonomies can be beneficial for organisations, the use of a central fixed classification scheme often becomes restrictive, belying the multidimensional nature of content, its subject matter and its users. Rather than a one-dimensional taxonomy, the multidimensional perspectives that metadata provide can be used to structure content in multiple ways enabling large data sets to be clustered around similar subject themes (Munkvold *et al.*, 2003). This approach to the use of metadata is referred to as faceted classification. Facets are orthogonal categories of metadata used to characterise content (Yee *et al.*, 2003; Hearst, 2006). Each facet represents one dimension but, in combination, multiple facets show several dimensions reflecting a broad range of user perspectives (Giess *et al.*, 2008; Tevan *et al.*, 2008). When applied in information retrieval, it combines the strengths of both free text search and structured navigation, and is increasingly applied in library databases and e-commerce websites (Hearst, 2006; Broughton and Slavic, 2007; Ben-Yitzhak *et al.*, 2008; Tevan *et al.*, 2008). The evolution of faceted classification included the development of the colon classification whereby subjects can be viewed and classified from five intrinsic perspectives: personality, matter, energy, space and time (Ranganathan, 1963). Further extensive work has since been carried out, expanding, modifying and refining these dimensions, a comprehensive account of which is provided by Giess and colleagues (2008). Stouffs and colleagues (2002) also advocate the use of semantically enriched attributes as facets to improve information retrieval within collaborative workspaces. As all metadata attributes can be used as facets, the challenge in using a faceted approach is identifying what attributes will be most appropriate to meet the organisation's needs.

Enterprise model

The enterprise model is a process map of activities based on the detailed analysis of the organisation, its business processes (including all of its operations, partners, supply chain and customer networks) and their interaction with information throughout the business's lifecycle (Munkvold *et al.*, 2006). It provides a clear view of the organisation's processes. This is particularly important because the basis of any well-defined IM strategy is a clear alignment with the organisation's operational context including its business strategy, processes, goals and culture (McNay, 2002; Gyampoh-Vidogah and Moreton, 2003). Thus the process used to create, store, retrieve, review, update and distribute content must be reviewed and analysed to provide a basis upon which a contextual IM strategy suitable to the needs of the organisation can be developed and deployed (Robson, 1994; Rockley *et al.*, 2003; Gottlieb, 2005; Paivarinta and Munkvold, 2005; Hamer, 2006). It is an inclusive process requiring significant input from all stakeholders (Reimer, 2002).

Systems and technology

The primary function of technology within a holistic approach is to facilitate the effective implementation of a predefined strategy. Selecting the tool is thus based on the needs identified and the solutions developed for the other two components (i.e. the enterprise model and the content model). The importance of this contextualisation is emphasised by Schaeffer (2002), who warns that procuring the wrong software can be worse than procuring none at all. Along with functionality and scalability, any technology employed must be user-friendly, intuitive, usable and secure, conforming to the organisational quality management regulations (Munkvold *et al.*, 2003). Information produced in often heterogeneous technology platforms will need to be shared, transferred, stored and managed independently of the applications with which it was produced. Thus technology implementations should enable content integration while remaining flexible and scalable enough to cater for increasing content volumes and new information products to emerge in time (Reimer, 2002; Ross, 2003; Paivarinta and Munkvold, 2005; Munkvold *et al.*, 2006).

Implementation and change management

A holistic approach may give rise to new or different ways of working that are alien to the end users within an organisation. Consequently, along with the procedural and technological components, training, guidelines, standards and change management are crucial to its successful implementation and maintenance (Munkvold *et al.*, 2003; Paivarinta and Munkvold, 2005; Hewlett Packard, 2007). Supporting these behavioural changes may take time and will require effort but will ensure that the approach to IM becomes embedded and thus sustainable (Davenport, 2000; Gyampoh-Vidogah and Moreton, 2003).

Chapter 3

Planning and Implementation of Effective Collaborative Working in Construction

Mark Shelbourn and Dino Bouchlaghem

Introduction

Project delivery in construction is highly dependent on the effectiveness of the teams put together to execute the project. In many cases this is a function of how well participants work collaboratively and how effective the communications infrastructure used by the team is. However, the full benefits of effective collaborative working in construction projects have yet to be fully realised. This is partly because of the inefficient use of information and communication technology (ICT) tools.

Significant efforts have been invested to develop tools and techniques that enable distributed teams of professionals to work collaboratively. Some of these systems were able to improve some aspects of collaborative working but did not address the dynamics of construction organisations, and their people, projects and processes, sufficiently. There are, however, isolated pockets of users that have fully embraced the potential of collaborative working within their businesses realising that the role of ICT has to be closely aligned with the people who use it and the overall goals of the business.

This chapter presents the findings from a project, Planning and Implementation of Effective Collaborative Working in Construction (PIECC), that aimed to develop a decision-making framework that enables organisations to fully integrate ICT and the associated people and business issues into their projects and within individual organisations. It reports on the processes used to develop the framework and then describes its content in detail.

Background to the PIECC project

The PIECC project recognised the need for a balanced approach to managing collaborative working in projects in order to take full advantage of the benefits that ICT can provide. For this, the sociotechnical and organisational aspects of ICT implementations need to be given full consideration.

The project focused on supporting strategic decision making by identifying areas where collaborative working can be improved by integrating organisational (business), project and users' needs. The needs of the construction industry formed the basis for the development of a framework that facilitates the strategic planning and implementation of effective collaboration supported with working policies and protocols. When carefully planned and based on informed decisions, it is believed that these policies and protocols will help organisations improve their collaborative working practices, achieve better benefits from it and maximise the use of tools and techniques that are currently commercially available. Particular attention has been given to the challenging requirements of distributed, heterogeneous and transient construction project teams together with the need to facilitate ubiquitous and serendipitous collaboration between team members and across all stages in the project delivery process. The remainder of this chapter describes how a balanced approach for effective collaboration is achieved through the development of a collaboration strategy.

PIECC project aims and objectives

The main aim of the research was to develop a strategic decision-making methodology that guides organisations in the planning for effective collaborative working practices and the implementation of suitable tools and techniques. The associated objectives were to:

- review the state of the art of collaborative working with a focus on both practices and technologies;
- conduct a requirements capture survey for collaborative working in construction at three different levels – organisational, project and end users – and identify key areas for improvements in collaborative working;
- develop a framework for the planning and implementation of effective collaborative working, taking into account both the organisational business process and the project lifecycle.

Methodology

In order to achieve the objectives of the research the following strategies and research methods were adopted:

- review current 'state-of-the-art' practice on collaborative working in the construction and other industries;
- establish current practice for collaborative working within organisations through focused field studies that included questionnaires, semi-structured interviews and detailed case studies to identify the

requirements for collaborative working and critical factors to be considered at organisational, project and users levels;
- using a 'develop–test–refine' strategy from action research, develop the framework and appropriate tools for the planning and implementation of effective collaboration.

The framework for effective collaboration considered the organisational culture, project process and user requirements for the implementation of collaborative tools and techniques (see Figure 3.1). Thus both 'soft' (i.e. organisational and cultural) and 'hard' (i.e. technological) concepts were considered and combined to achieve the objectives of the research. Lessons from past research initiatives also suggested that the combined approach of 'soft' and 'hard' issues was the most appropriate approach to be used.

Key findings

Participants in collaboration ventures usually come together to build capacity and complete a set of tasks that no one single organisation can undertake in isolation. Good collaboration can eliminate fragmentation, duplication and distrust. This is achieved by using the resources available effectively, sharing project risk across multiple domains, and enhancing staff and organisational motivation.

It is important to note that no two collaboration ventures will progress in exactly the same way or within the same time frame. The PIECC philosophy was based on the principle that each collaboration endeavour should

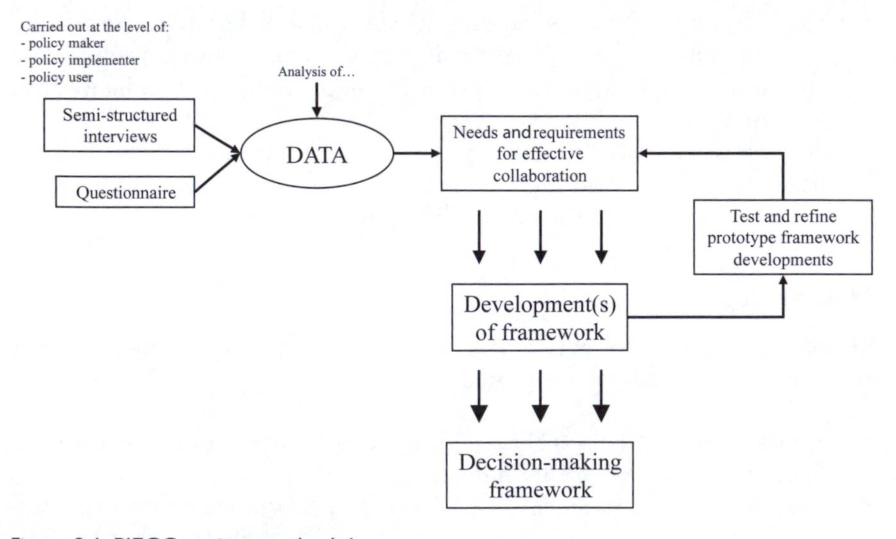

Figure 3.1 PIECC project methodology.

be managed in a way that is aligned with a unique set of circumstances and composition.

It is essential to allow the collaborating stakeholders to take the necessary time from routine responsibilities in order to meet and interact with one another so that trust and respect on an individual level can be established. However, these interactions will inevitably lead to the drawbacks naturally associated with the systemic change associated with the resulting new ways of working. Therefore it is important to keep in mind that the most important overarching principle of effective collaboration is that working collaboratively often means new ways of working (at least initially) for those involved.

The findings indicated that effective collaboration is achievable only through the implementation of a well-balanced collaboration strategy that does not rely solely on sophisticated ICT systems. It was confirmed that there is little guidance and support available to effectively implement and manage collaborative project environments. Therefore the decision-making framework developed as part of the PIECC project and described in this chapter consists of a new integrated strategy that enables full benefits to be realised from a more targeted use of ICT that is better aligned with the organisation's people and business processes.

Development of the PIECC framework

The requirements for effective collaboration support were established through focused field work as illustrated in Figure 3.1 and involving six large contractors and engineering consultants. The findings highlighted the need for an integrated framework that brings together hard and soft aspects of collaboration in the form of an intuitive and flexible tool that is applicable in different contexts and situations at both business and project levels. It should build on other accepted industry standards and practices in order to encourage uptake and adoption. The field work established that the model should include the following components:

- *Processes* that enable participants to agree a common vision and priorities for the collaboration with a clear route map for how the project is to proceed and an ongoing review of progress against vision and objectives. This should be supported with clear procedures for managing interactions and communications to promote trust, together with strong leadership from a 'collaboration champion'.
- *Appropriate standards and protocols* that facilitate interoperability between different software and systems to avoid the difficulties associated with having to adopt different applications on different projects. Protocols for communication and the sharing of information should

also form part of the collaboration strategy to promote trust and build good working relationships between collaboration participants.

- *Tools and techniques* to support the processes and tasks involved in the planning and implementation of collaborative working with ongoing monitoring and feedback mechanisms. Whenever applicable, examples of good practice and case study material that shows tangible business benefits of effective collaborative working with evidence should be promoted and disseminated as part of the collaboration strategy.

Having established the requirements, the work moved into developing the decision-making framework for effective collaborative working. Close interactions with the industrial partners and detailed examination of relevant literature revealed that a 'develop–test–refine' approach was the most appropriate for this stage.

Four stages were identified as key to a collaborative working strategy; these are shown in Figure 3.2. The first iteration in the development of the framework built on this initial concept, integrating the six key factors for effective collaboration with the three key strategic areas described in Chapter 2.

Feedback from the industrial partners on the initial version of the framework indicated that a more process-centric view is needed. This issue was addressed in the second iteration of the framework development. This cycle began with a redesign to introduce a greater process focus, splitting it into a number of stages: the first aimed to bring together the different organisations involved in the collaboration to 'align their business strategies'; the second stage required the 'collaboration champion' or 'team' to define the three strategies for effective collaboration (i.e. 'people', 'project' and 'technology' strategies); once these are defined the collaboration can begin and is continuously monitored to make sure that all participants are working in

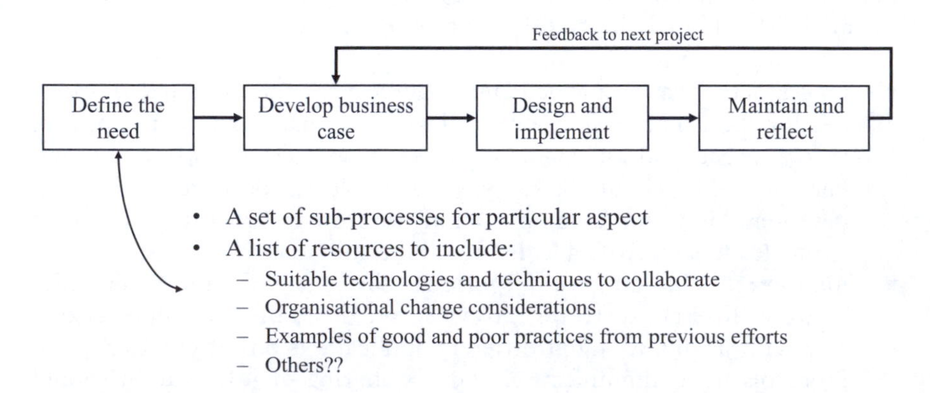

Figure 3.2 Initial thoughts of the decision-making framework.

the agreed ways; finally, feedback on how to improve current practices is captured and disseminated to the business.

The end of the second iteration of the framework development cycle revealed that the supporting information and how it is presented was of vital importance to the success of the framework. The feedback at this stage pointed out that this should be made more 'practitioner-friendly' as this is the main audience of the framework.

The industry steering group commented that version three adequately represented the key processes and supporting information; however, the framework required further development to make it more user-friendly with more guidance on key decisions and when they are made.

During one of the project workshops, the research team and industrial partners created a better graphical representation that shows how the processes are linked together in a more usable format. After further consultations the final version of the framework was completed; this is shown in Figure 3.3.

The PIECC framework

The framework contains four distinct sections that structure the planning and implementation of the collaboration: set up the business strategy; define a collaboration brief; plan a solution to fulfil that brief; and then implement that solution. The collaborators do this by working from the top of each section, following the processes to develop the strategies with an increasing level of detail as they move from left to right through the framework. Before progressing from one section of the collaboration to the next and depending on the outcomes of the different processes within that section, a decision is needed on whether it is feasible to continue. If the decision is 'no' then the team members need to revisit that section to try and remove any barriers or problem areas preventing the collaboration from proceeding. If it is established that the collaboration can continue then the collaborators move into the next section of the framework to carry on with the planning for the collaboration. The following sections describe the different processes of the framework; they should be read in conjunction with Figure 3.3.

Business strategy

Establish need to work collaboratively

A common reason for unsuccessful collaboration is the lack of sufficient senior management backing and sponsorship. To establish the need for the strategic planning for ICT and people, large companies may wish to conduct a feasibility study, whereas smaller companies might do a cost estimate

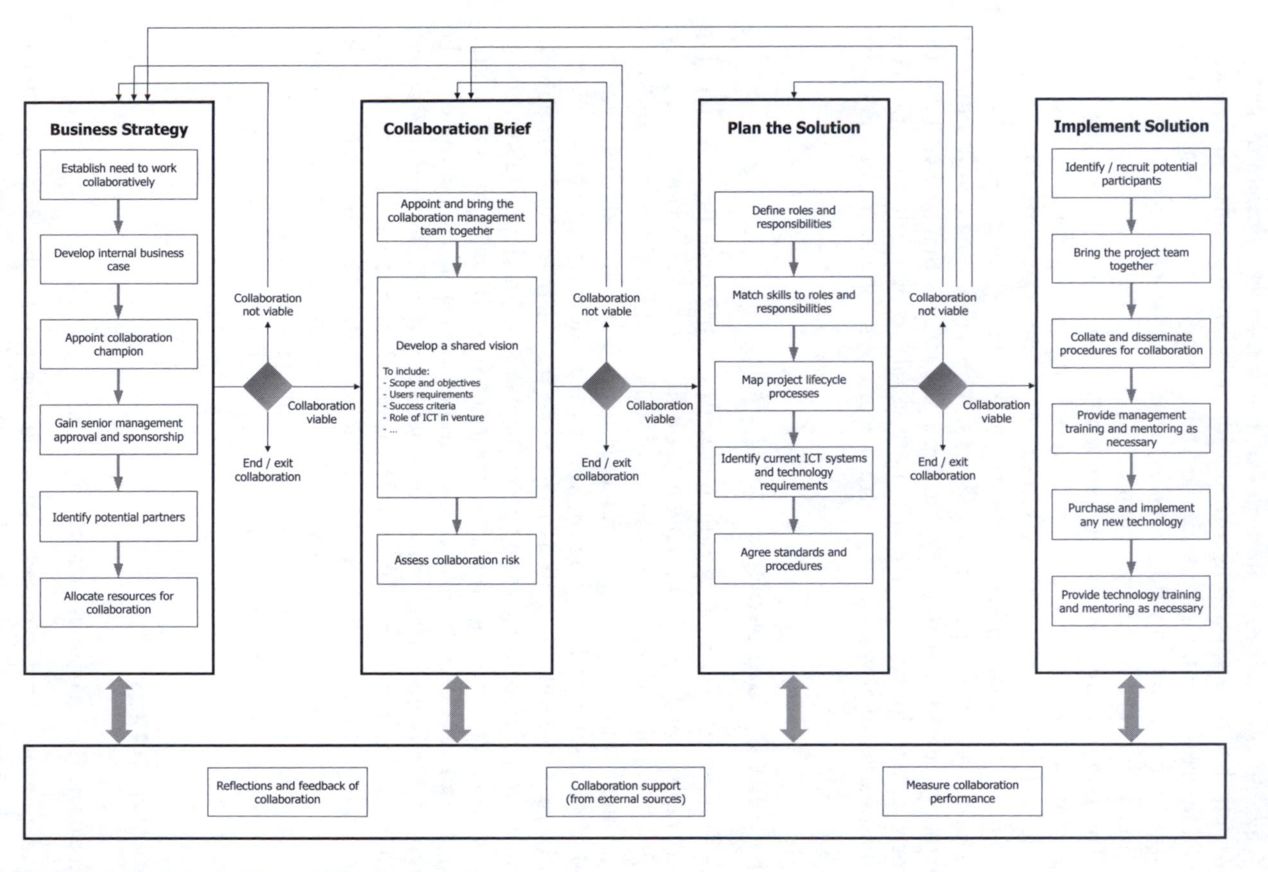

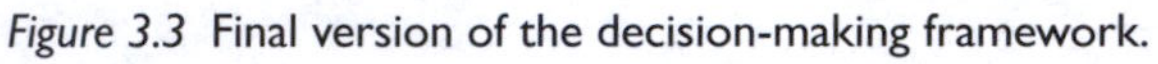

Figure 3.3 Final version of the decision-making framework.

of the process. In either case, it is advisable to gain support from a senior manager who would also sponsor the whole collaborative venture from its inception to implementation. In certain circumstances the client will specify requirements that dictate to individual organisations or an established project team on working collaboratively. In these cases it is vitally important to satisfy these requirements.

The person who is proposing to the organisation that working collaboratively is beneficial should lead the development of data and information to highlight the need for doing so. They may be able to do this as an individual or as part of a team. External support from consultants may be needed if the proposed collaboration is large enough.

To enable the clear establishment of the need to work collaboratively, the project leader in each organisation may conduct a feasibility study to make a clear case to senior management. Feasibility studies are designed to provide an overview of the 'make or break' issues related to a business idea, in this case working more collaboratively, and whether or not it makes sense for the business. The list below highlights some of the questions to consider in the four key aspects of conducting a feasibility study.

MARKET ISSUES

- Can we afford not to adopt collaborative working procedures?
- Is there a (current or projected) demand for adopting collaborative working procedures?
- What competition exists in the market?
- Are there any products/procedures available to meet the requirements – and what is their cost?
- How easy are these products to use?
- What is the availability in the locations?

ORGANISATIONAL ISSUES

- How do the different company cultures, ethos, mindset and policies affect the collaboration?
- Who will serve as senior personnel on the collaboration board?
- What qualifications are needed to manage the collaboration?
- Who will manage the collaboration?
- What other staffing needs does the collaboration have? How will these needs change over the next 2–3 years?
- What key experience does the organisation have that can be built on?
- What areas have a lack of skills and experience that needs addressing?
- What are the key processes that will be needed to ensure success of the collaboration (both own organisation and others involved)?
- What are the potential user-training requirements?

- What are the controls that need to be in place?
- How will performance/compliance be dealt with?

TECHNICAL ISSUES

- What are the technology needs for the proposed collaboration?
- What other equipment does the proposed collaboration need?
- Where will you obtain this technology and equipment?
- When can you get the necessary equipment?
- How much will the technology and equipment cost?
- What are the implications of ownership of technology and equipment (including maintenance and ongoing support)?
- Integration with existing systems – is it needed, how will it be achieved, what will it cost, what is the business case for integration?
- How does the technology fit with existing technology – compatibility?

FINANCIAL ISSUES

- Start-up costs – these are the costs incurred in starting up a new collaboration: training course preparation and attendance, preparation of literature material and printing costs, meeting and travelling expenses as necessary.
- Operating costs, such as rent, utilities, support and maintenance of technology and systems, ongoing training and wages, that are incurred in the everyday operation of the collaboration.
- Revenue projections – how will you price the collaboration service? Assess what the estimated monthly revenue may be.
- Sources of financing – research potential sources of borrowing to finance the proposed collaboration.
- Profitability analysis – the 'bottom line' for the collaboration. Will the collaboration bring enough income to cover operating expenses? Will it break even? Is there anything you can do to improve the bottom line?
- What is the total cost of ownership, including pay-back period, whole lifecycle costs, sunk costs and opportunity costs?

Develop internal business case

The business case presents a view of the collaboration venture to senior management in the individual (potential) partner organisations. It should provide the basis for financial justification and return on investment information. It should be seen as a critical component to win support for the venture, and provide a base from which to make the case for preparing the way the organisation will work. The person(s) suggesting a move towards a collaborative venture should lead the development of the business case.

They should highlight key personnel to include as part of the case to show support. The case should be used to:

- communicate the details of the venture to others;
- establish a method for measuring success (also see 'Develop a shared vision' in the 'Collaboration brief' section);
- receive funding approval for the collaborative venture.

The business case should tell the collaborative venture's 'story' in straightforward, easy to understand language. If done correctly, the business case will provide compelling justification for a change in working practices in the organisation (i.e. collaborative working) by outlining, at a high level in the organisation, what is wrong with the current situation, a potential solution and its possible impacts. The business case should answer the following questions:

- Why are we engaging in this venture?
- What is the venture about?
- What is our solution to the business problem?
- How does this solution address the key (individual) business issues?
- How much will it cost?
- How will the business benefit?
- What is the return on investment and pay-back period?
- What are the risks of the venture?
- What are the risks of not engaging in the venture?
- How will success be measured (see 'Measure collaboration performance' process)?
- Are there any other alternatives?

Appoint collaboration champion

Organisations that successfully plan and implement effective collaboration use a key employee or 'champion' who drives the process with energy and commitment. This person must have an understanding of the potential and role of ICT systems, people and business processes, along with detailed knowledge of their organisation's processes and practices. This champion is often also the project manager of the collaborative effort. In smaller firms, the champion and the sponsor could be the same person.

Senior management in each organisation must appoint a suitable champion to represent their organisation in the collaborative venture. This champion should have a proven track record as an effective leader. They must demonstrate knowledge and experience in the collaborative working arena, as they will be looked to for guidance. They should demonstrate that they have authority and most definitely have experience with collaborative working. As the leader they must be engaged in the collaboration process – actively, visibly and continuously. The role the champion plays and the behaviours they demonstrate can often build up or tear down the collaboration.

Gain senior management approval and sponsorship

The gaining of support from senior managers or directors of the organisation is crucial for the collaboration to move forward. The senior managers or directors must have the power to influence strategic decisions regarding the business direction of the organisation. Their role in the planning and implementation process for collaboration can be summarised as:

- ensuring that any collaboration activity is integrated with the business decision-making process of the organisation;
- providing sufficient resources to plan and implement the collaboration;
- maintaining an overview of the collaboration and ensuring that changing business processes are incorporated into the organisation.

The collaboration champion (and, when necessary, the project team) needs to present the need to work collaboratively to their organisation's senior management. In certain situations the presentation will also need to be made to the client of the project to ensure that their support is gained for the collaboration.

A clear and coherent business case has already been put together and it is vitally important to present it to senior managers in the organisation and/or the client, whichever is the potential sponsor of the project. Therefore a meeting should be scheduled to present the case for collaboration.

Identify potential partners

Some organisations will have well-defined policies, procedures and practices whereas others do not. Some organisations identify people from within whereas others hire expertise and experience from the outside. In some organisations people are valued for their competence, expertise or creativity whereas in others people are valued for their seniority. Some organisations behave in a hierarchical way, with clear regard for title, seniority and/or authority, whereas others prefer democracy or anarchy as the primary form of governance.

It is important to note that potential partners and their associated personnel will carry to the collaborative venture these constructs of the organisation they represent. Failure to deal explicitly with these issues and reconcile the differences will result in individuals and organisations relying on their own experiences to determine how the venture will operate for them.

It is the role of senior management to provide preferred partners to approach to join the collaboration venture. It may also be specified by the client that certain organisations will have to work together in a potential collaboration venture.

Within this process it is important to consider:

- what is needed in another collaborating organisation;
- what criteria are used to decide that an organisation is suitable for the venture.

Many organisations looking for potential partners consider technical or functional expertise first and foremost. Many also aim for the 'best' to be part of their team in terms of knowledge, skills, reputation, accomplishments and interests.

Some organisations will have well-defined policies, procedures and practices whereas others do not. It is important to consider these differences in highlighting potential partners for the venture. Also, regardless of the type of collaborative venture you are planning, you must concern yourself with having the right people involved. 'Right' in terms of:

- the abilities of the individuals to perform the required *functions*;
- the way (*form*) in which the individuals do their work; and
- whether and how each individual *fits* as part of the venture.

See also the 'Identify/recruit potential participants' in the 'Implement solution' section of the framework for more details.

Allocate resources for collaboration

The performance of every business is largely dependent on how effectively senior managers plan for and allocate the finite resources to meet their strategic and tactical objectives. A company that can allocate resources where and when they are needed most, and quickly reallocate resources to address changing business conditions, can successfully capitalise on emerging opportunities that collaborative working provides.

Providing appropriate resources for the collaboration effort will convey a positive message, to other collaborating organisations in the venture, of the commitment towards ensuring a successful venture. Senior members within the collaborating organisations must therefore have the power to allocate the appropriate amount of resources to the collaborative efforts. Where and how this resource is allocated must be determined by the collaboration team under the guidance of the collaboration champion.

Organisations need to plan to ensure that the necessary resources are in place to support the delivery and management of tasks to be completed in the collaborative venture. This will include allocating a budget for:

- engaging the stakeholders who will be involved in the project;
- the purchase of hardware and software licences for the development and delivery of tasks;
- the professional development of staff involved in the collaborative venture;
- training (systems and processes) and development of collaboration procedures and processes.

Collaboration brief

Appoint and bring the collaboration management team together

Appoint a management team that represents a cross-section of the collaboration participants' expertise. These staff need to be experienced personnel in each of the participating organisations and have a particularly clear understanding of the principles of collaborative working and the potential of ICT and its effect on collaboration.

In most collaborative relationships it is not always possible to have every participant directly involved in every decision. Size and complexity of the venture dictate the need to create some form of government. The government is represented here as a *management team* to be formed by the collaboration champion to provide leadership for the venture. This team should comprise a manageable number of key representatives from the principal organisations involved in the venture, ensuring some organisational or contractual arrangements are in place for every participant in the venture. Key responsibilities for the management team include direction setting, operational coordination and problem escalation.

Whether a collaborative venture is large or small, local or global, efforts should be made to help participants to interact with each other. This will help develop trust and respect among them. Depending on previous collaboration history, and how different professionals deal with trust and respect, there are a number of methods that can help this process. They include:

- a review of professional résumés;
- facilitated discussions;
- personality profiling and self-discovery tools such as the Myers–Briggs Type Indicator (see http://www.cpp.com/products/mbti/index.asp) and Strengths Deployment Inventory (see: http://www.personalstrengths.com/catalog_sdi.htm#top);
- social events;
- team-building sessions.

Develop a shared vision

An important stage in the planning for a successful and effective collaboration is the development of a shared vision. Experience shows that, when a shared vision is agreed, other factors fall in place more easily. Such a vision must be large enough to include the scope and objectives; success criteria; and an initial decision on the role of ICT in the venture. Vision regarding the desired outcome:

- defines the playing field and the rules of the game at a high level; and
- enables all participants to align their own activities and behaviours to preserve core beliefs and values and make progress towards the overall goal.

A shared vision provides meaning, direction, focus, boundaries and guidance to all those in the collaborative venture. This vision should become a basis for a common understanding to be developed for the collaboration.

Setting objectives and establishing clear priorities helps to solidify the purpose and manage everyone's expectations of what will be accomplished, by whom and by when. However big or small, measurable or fuzzy, long-term or immediate, the more participants can do to instil a common understanding of the criteria for a successful collaborative venture, the more likely it is that all involved will experience successful collaboration. The criteria must be finite, sharp, vibrant, crisp, concise and compelling.

All participants need to have the same 'vision' in their minds. The collaboration champion (or team) needs to know how the individual activities contribute to reaching the required targets, and when and whether progress is being completed towards these targets. However it is labelled, the vision for success must encapsulate the fundamental reason for being; guiding principles; values and beliefs; the overall goal for the venture; a high-level measure of success; and initial decisions on what role ICT will play in the venture. The success criteria should be:

- well defined;
- measurable (relating to 'measure collaboration performance' process);
- related to individual activities.

The collaboration champion – with the help of their management team where appropriate – must lead the development of the vision for the collaboration. They must invite as many of the stakeholders from the supply chain as possible to participate in the developments. A vision is composed of two parts: philosophy and picture. Both are necessary; neither is sufficient on its own.

- *Philosophy* includes guiding principles, values and beliefs that are held dear by the group, along with an expression of the group's fundamental motivation or primary reason for being. These capture participants' hearts and provide a level of moral guidance for participants' plans, activities and behaviours.
- *Picture* includes a clear and compelling overall goal along with a vibrant, tangible measure of success. When it is cogent, concise and compelling this picture serves as a unifying focal point for the effort.

A Memorandum of Understanding (MoU) is a useful way of documenting the shared vision, and in establishing the basis for the sense of seriousness of partner organisations and their intentions. The MoU creates a non-legal bond to a particular course of action to help focus negotiators' minds on the issues and to record the 'meeting of minds' to that point. It can also:

- establish a framework for future negotiations;
- record oral understandings so as to help prevent misunderstandings occurring later;
- serve as a memory aid for the draftsperson when later drawing up any contractual documents;
- be used as evidence to a third party, such as the client, that a contemplated venture is viable; or
- be used for publicity purposes to indicate a particular course of action is under way.

Whilst developing the shared vision it is important to establish the role of ICT in delivering the objectives. The role of ICT cannot be decided after setting the project strategy since there may be situations where subsequent activities when establishing current systems and technology needs reveal that the shared vision cannot be delivered. It is also vitally important, at this early stage, to include a coherent agreement of how such technologies are to be used in the collaborative venture.

A common understanding of how ICT should run alongside the collaboration strategy is required, as a technology may be necessary to support the project's collaborative processes (such as videoconferencing to support face-to-face communication and teambuilding among disparately located teams) as well as supporting general functions of the project team. At the time the role of ICT is decided it is also important to determine the skills within the project team and organisations.

It is recommended that the project scope and objectives, project team roles and skills, and the role of ICT in delivering the project be established in parallel. Each must be linked back to the project vision and scope; for example, if the project vision covers long-term whole-life objectives then the roles of project team members and the role of ICT in the project may be different from a traditional design and build project.

Assess collaboration risks

Risk assessment is a method of identifying, analysing, communicating and controlling risks associated with any activity, function or process in the collaboration. Assessments can take different approaches depending on the purpose and scope of the information or data used. Some assessments look back to try to assess effects after an event, such as an accident on site. Other assessments, as in this case, look ahead to predict what the effects will be before a (new) collaboration begins.

A suitably qualified member of the management team should be identified to be responsible for the formation of a group (if necessary) and the management of the risk assessment process. This person must have a proven track record of conducting such assessments as well as a full understanding of the new ways of working associated with working collaboratively. If no such person is available then external support should be sought.

Using the model shown in Figure 3.4, a detailed risk assessment can be conducted for many of the processes and procedures to be employed during the lifecycle of the collaboration. The model is made up of four stages.

To aid the risk assessor (and their team) to determine risks for the collaboration, a useful matrix (shown in Table 3.1) may be used. The matrix matches the 'probability' of a risk occurring against the 'severity' of the risk. This matrix provides a useful tool in determining risk factors for the processes and procedures to be used in the collaborative venture.

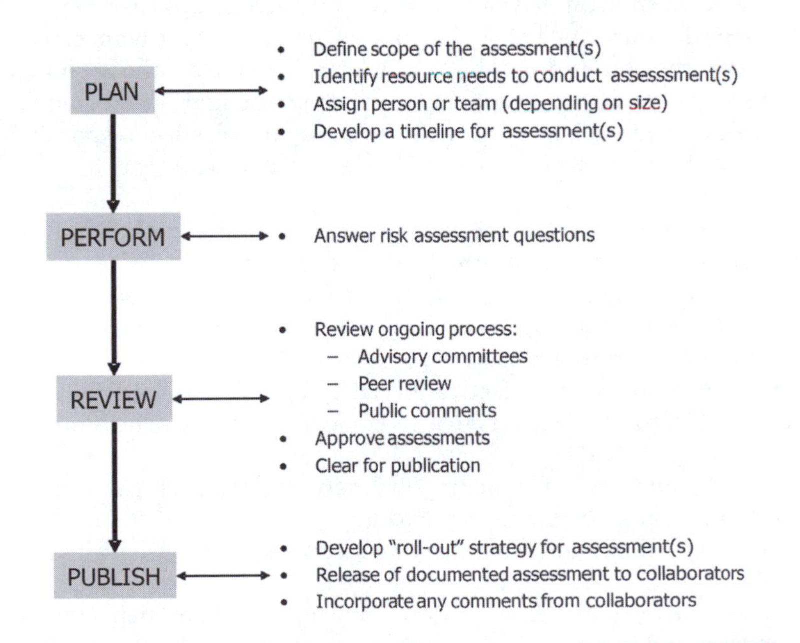

Figure 3.4 Four-stage process for conducting risk assessments (adapted from http://www.cfsan.fda.gov/~dms/rafw-toc.html).

Table 3.1 A matrix for assessing risks

Probability				
	High			
	Medium			
	Low			
		Low	Medium	High
			Severity	

The 'Be Collaborative Contract' has a whole section dedicated to determining risk. It can be found at http://www.bcc.beonline.co.uk/risk/risk_1.html

Plan the solution

Define roles and responsibilities

Consistent with good business practice, it is well established that people perform better when they know what is expected of them. In addition, in a collaborative venture it is important for all to know what they can and should expect from each other. This provides an opportunity to match responsibilities with authority, something that can be lacking within organisations, never mind in cross-functional or multicompany teams.

Trust between team players is enhanced if there is transparency in roles and responsibilities, and particularly if each member of the team can understand what the others are doing and whether they have the authority for such tasks. Efficiency and economy of effort is essential for all members of a collaborative team, and by relating the clarity of roles and responsibilities to the collaborative team's processes one can eliminate 'gaps' and avoid duplication.

It is the responsibility of the collaboration champion with help from the management team to define the roles and give appropriate responsibilities to participating members from the different organisations involved in the collaboration. They are also charged with the monitoring of progress against those responsibilities given.

Many techniques are effective in clarifying roles, responsibilities and accountability for participants in a collaborative venture. These include:

- abbreviated forms of standard job description for each participant
- ongoing dialogue among participants
- documented minutes from working meetings articulating who is doing what
- roles and responsibilities matrix, indicating for key deliverables and activities who:

- is accountable for the outcome/drives the process
- contributes to the work/supports the effort
- MUST be kept informed along the way
- has decision-making authority/approval authority.

A typical 'management' structure for a collaborative venture is shown in Figure 3.5. A definition of the key roles for collaboration is shown in Table 3.2.

A popular method used to define roles and responsibilities is to use RACI (sometimes called RASCI) charts. RA(S)CI is an abbreviation for:

- R = *Responsible* – owns the problem/project
- A = to whom 'R' is *Accountable* – who must sign off (*Approve*) on work before it is effective
- (S = can be *Supportive* – can provide resources or can play a supporting role in implementation)
- C = to be *Consulted* – has information and/or capability necessary to complete the work
- I = to be *Informed* – must be notified of results, but need not be consulted.

An example of a RACI/RASCI chart is shown as Table 3.3.

Match skills to roles and responsibilities

Having determined the roles and responsibilities required for the collaborative venture the next stage is to break these roles down into individual tasks. These tasks will then be used to define the skills needed and then identify the right staff with the relevant skills. An important consideration here is the role of ICT in the collaborative venture where the identification of the people skills and the technologies to be implemented needs to be

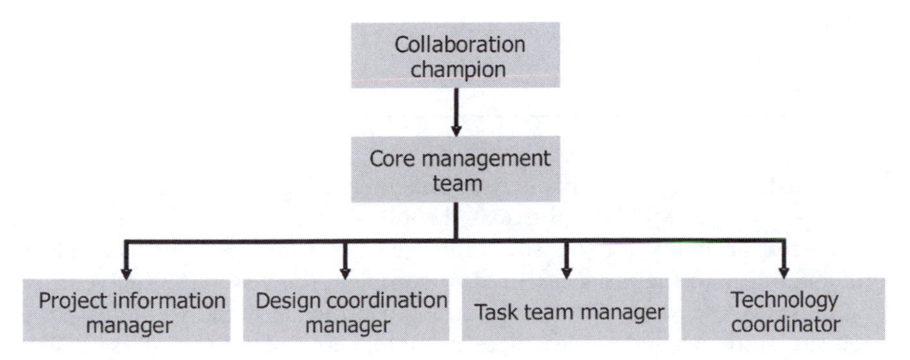

Figure 3.5 Typical management structure for a collaborative venture.

Table 3.2 Typical roles for a collaboration venture

Role	*Description*	*Responsible for*
Collaboration champion	The leader of the collaborative venture	The decision making and smooth running of the collaboration Solving problems and difficulties as they arise in the day-to-day activities of the venture
Core management team member	A member of the core team that manages the collaborative venture	Aiding the collaboration champion in managing the collaborative venture
Project information manager	Manages all documentation in the project	Ensuring that all data and information used in the project is in the correct form, and the most up-to-date is being used
Design coordination manager	Organises all the design-led personnel in the project	Ensuring the design and construction teams are working effectively throughout the lifetime of the project
Task team manager	Manages the different tasks in the venture	Ensuring all production of outputs relate to a specific task
Technology coordinator	Manages all technology in the venture	Ensuring that the technology used in the venture is compatible, and that all users are up to speed on its use
Collaboration member	Any person who is to be a part of the collaborative venture	Ensuring that they follow the standards and procedures agreed for the collaboration

Table 3.3 An example of a RACI/RASCI chart

	Collaboration champion	*Project information manager*	*Design coordination manager*	*Task team manager*	*Technology coordinator*
Activity 1	R		A		
Activity 2	A	R		R	C
Activity 3			I		I
Activity 4	R		A	C	
Activity 5	A	R			S

conducted concurrently. This is to avoid having highly skilled individuals use basic tools, or highly sophisticated tools used by unskilled personnel.

The responsibility for this should be given to someone who has a clear understanding of the tasks required to complete the vision for the venture and must be able to breakdown the tasks in sufficient detail to identify the requisite skills needed to complete it. Before any matching of skills can take place, detailed job and people specifications need to be made available.

Tables 3.4 and 3.5 define the main elements of a job description and a person specification.

Map project lifecycle processes

Working collaboratively often involves a substantial financial outlay that is not always covered in 'normal' project finances, therefore there is little benefit in searching or developing and implementing ICT systems that simply automate inefficient, conflicting and unnecessary business activities.

Well-defined processes form a critical part of developing and maintaining effective collaboration. Any lack of clarity about the processes and systems to be used by participating organisations may lead to coordination problems. A purpose is only academic until the 'how to' is determined. Similarly, whether and how to establish trust and confidence in others is dependent on seeing and experiencing how things are going to work in the venture. Knowing who is doing what for whom, when and where is critical in order to be productive and positively contribute to the collaborative venture. It is

Table 3.4 Job description details

Job description	A job description should be carefully prepared to identify the key duties and responsibilities to be undertaken. It should be written in clear, straightforward and gender-free language and should avoid gender stereotyping of jobs
Job title	
Job purpose	To accurately reflect the nature of the job
Job duties Sub-set listed here	This should not merely list tasks, but should emphasise the objectives of the job. The purpose of each duty should be clearly defined and specific terms should be used. Vague expressions such as 'administration' should be avoided. Although the duties section should be specific it should not be too prescriptive or restrictive. The nature of the duties will vary over time without affecting the overall job purpose, therefore the job description should allow for flexibility of approach. The job description should not emphasise aspects of the job which may discourage certain groups of applicants when, in reality, such aspects are of minor importance, nor should it contain words that imply that most of the people currently doing the job are predominantly of one particular gender
Special conditions	This section should be used to identify aspects of the job which can be regarded as unusual, such as non-standard working hours, restrictions on holiday, call-out liability, etc.
Organisational responsibility	This section identifies the functional relationship of the post, showing whom the member of the venture is responsible to and at what level, as well as the number and type of staff that they in turn will supervise (if required)

Table 3.5 Person specification details

Person specification	A person specification identifies the critical attributes required in a candidate if he/she is to be capable of carrying out those duties and responsibilities to a satisfactory standard. The criteria contained in the person specification should be strictly relevant to the requirements of the job and must be clearly justifiable in terms of the ability to perform the duties of that job *Essential:* Those criteria that an applicant must possess to be able to the job. If someone applies for a job, but does not have one of the essential criteria on the spec, they should not be offered the position. This demonstrates how important getting the spec accurate is *Desirable:* Those criteria which would be advantageous, but are not critical to the post
Experience	Define the work experience that is necessary for a person to have before the job in question can be performed. Avoid specifying an arbitrary number of years' experience as essential, as the quality and range of experience is more important than the length of experience
Skills and abilities	Define the practical skills and abilities that are required to perform the job. Be as specific as you can. For example, 'Good computer skills' does not really define what actual skills are required, whereas 'computer skills sufficient to be able to produce complex documents and statistics' is much easier to define and measure
Qualifications	State the minimum educational or vocational qualifications required
Training	Outline the practical training which the post holder will have to complete in order to undertake the job satisfactorily
Other	This section will indicate particular characteristics the post holder should possess to carry out the duties of the position. For some jobs there will be specific requirements; these should be listed

important to build into project schedules time to build relationships with fellow collaborators. This will begin the process of building trust in the collaboration, an essential ingredient for success in the collaboration.

The collaboration champion and management team must administer the decision-making process of how the mapping of rules and procedures onto the scope and objectives of the collaborative venture is conducted. They must do this by involving key representatives from all collaborating organisations and seeking their views of how they wish to work.

As the old saying goes, success is in the details. Whether participants create and maintain effective collaboration depends in part on how they work together. It depends on how well processes are articulated and adhered to throughout the venture, including the systematic series of actions participants use to achieve some end and the 'rules' they follow. However, it is best to model the business not in too much detail, but just sufficiently well to capture the key activities of each organisation. Once each organisation has modelled its working practices, a new model can be created to incorporate shared processes that already exist and then used as an example and starting point to merge other areas where shared processes are required.

To ensure effective collaboration, participants must take time to articulate how specific collaborative relationships will operate. Successful collaborators in many different industries and situations repeatedly point to the following as fundamental to their success:

- how to get along with each other;
- how to solve problems and resolve conflicts;
- how to stay focused on the important aspects of the work;
- how to run meetings well;
- how to make and follow through decisions;
- how to divide up the work.

The collaboration champion must organise a meeting to fully understand how each organisation functions, by modelling each organisation in a graphical form. A small to medium-sized organisation can create a 'hierarchy of activities' diagram, as shown in Figure 3.6. Larger organisations can use more sophisticated modelling techniques such as data flow diagrams and workflow diagrams.

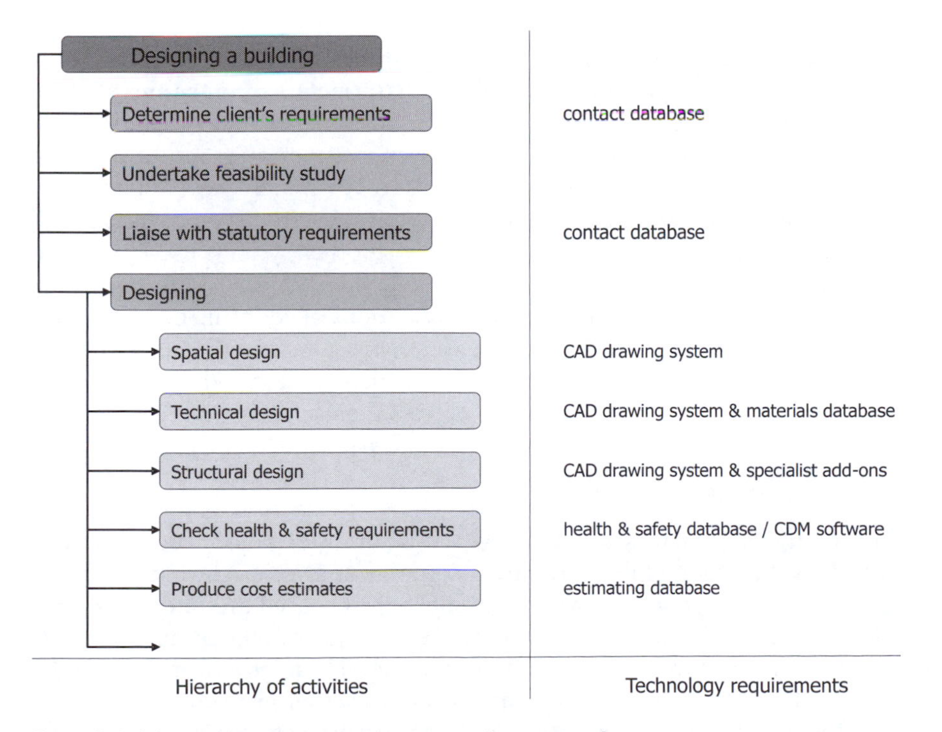

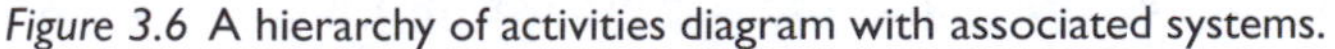
Figure 3.6 A hierarchy of activities diagram with associated systems.

Identify current ICT systems and technology requirements

Having established ICT roles within the collaborative venture in the development of the shared vision, it is important to establish what technology is currently in place, as there may be opportunities to adopt a particular type of system or system supplier across the collaboration venture. Alternatively, the gap between existing technologies and technology needs may be significant and the cost of upgrading existing technologies, or purchasing new ones, prohibitively high.

The main aim here is to manage collaboration around agreed standards and procedures, it is therefore necessary to fully understand the results of a technology audit. It may not be as simple as just assuming that two compatible systems or tools will facilitate collaboration if two companies use those systems and have staff of varying skill levels.

Having established what ICT systems (software and hardware) are currently in place, the next step is to establish what is required in order to achieve the collaboration objectives taking into consideration the characteristics of the project itself.

The collaboration champion and management team should conduct the survey to determine current ICT systems and technologies. Analysis of this survey should reveal the requirements for systems, technology upgrades and purchases.

There are alternative ways of establishing what ICT systems and technologies are in place within project team member companies. When undertaking any audit it is important to realise that many companies will have multiple versions of any software systems operating across projects and that a single system can be used in different ways. So, when determining what is in place, it is also important to establish the type of usage of a particular system, and if it is compatible with other systems used by other project partners.

It is also important to consider the implications of adopting new systems; these range from the cost of purchase and maintenance to the need for personnel training. There may be situations in which the cost of the ICT systems required is so high that the project objectives themselves are affected.

Agree standards and procedures

One of the key factors for effective collaboration is good communication. It is therefore vital to lay down some ground rules so that the basic requirements for communication and mechanisms used are understood by all project participants. It is also important to ensure that communication occurs in a structured and consistent manner. These ground rules should consist of information exchange and communication protocols and standards for structuring, representing and sharing project information. These protocols and standards can vary from one organisation to another, and failure to manage these differences can lead to inefficiencies during the project. Standards should cover the production of information within ICT systems, the formatting and structuring of that information, and procedures for the exchange and reuse.

Often information reuse is neglected during the selection and implementation of ICT systems; team members only consider their own roles in generating and exchanging information. The reuse of information is one of the major areas for efficiency savings through effective collaborative working; however, it does require a level of trust about the quality of the information being provided and the ways in which it is reused. This trust is a function of both the project contract arrangements and the working relationships between team members.

It is always useful to establish what standards and protocols the project team members already have in place. There is usually a good deal of commonality between them so, in theory, agreeing on some of the core principles should be reasonably straightforward. However, in practice it is often more difficult to achieve as most team members are usually confident of the benefit of their own processes, inflexible to change, or hamstrung by their internal quality assurance procedures. It is important to get team members to understand that some compromise is necessary if a common set of standards and protocols is to be agreed by all.

The development and/or implementation of standards and protocols is best as a robust, well-defined process undertaken as part of the wider project management processes and not as a discrete ICT-related activity. Just as the ICT strategy supports the project strategy, the information standards and protocols should support project processes.

Implement solution

Identify/recruit potential participants

A major factor that influences successful collaborative ventures is the participants' abilities to both accomplish tasks and manage relationships with others. As previously described in the 'identify potential partners' activity under the business strategy area, it is vitally important that the 'right' people are involved in the collaborative venture; 'right' in terms of:

- the abilities of the individuals to perform the required *functions*;
- the way (*form*) in which the individuals do their work; and
- whether and how each individual *fits* as part of the collaborative venture.

The collaboration champion and the core management team must be responsible for choosing the collaboration members. They need to create the right mix of people as individuals have different sets of qualities that define their personality and hence their way of working and interacting with other professionals.

Using the information from the 'define roles and responsibilities' and 'match skills to roles and responsibilities' processes, people to be associated with the venture can now be chosen. This should be done by examining the skills of existing personnel in the organisations. If the right individuals are not part of the organisation then external appointments can be considered. To help examine skills and behaviours, the following checklist can be useful in searching for individuals who may be effective in a collaborative venture:

- functional competence – appropriate level for the work to be done;
- fluency in the 'language' of the venture – proficiency with jargon, acronyms, colloquialisms unique to the venture;
- clear role to play on the team – specific accountabilities and deliverables;
- personal passion for the work – how much a person likes their work and its impact on others;
- service mentality – the level to which a person wants to support and collaborate with others;
- cultural fit;
- 'welcome' factor – how well a person is accepted by others in a group.

These should be used in conjunction with the job and person specifications developed earlier in planning the solution activities. However, the best way to recruit people is to do as little recruiting as possible. This means keeping as much as possible of the talent that you already have. Turnover of staff has real costs that can be as high as 1.5 times the departing person's salary, not to mention the effects on workload and the morale of the collaboration.

Other publications to consider reading include:

- 'Selecting the Team' publication from the Construction Industry Council (http://www.cic.org.uk/services/publications.shtml);
- 'Managing People on Construction Projects' publication from the European Construction Institute (http://www.eci-online.org/pages/publications-new.html).

Bring the project team together

The team brought together to undertake the shared project must be able to work in a trustworthy manner to build competence, develop interpersonal relationships and perform collectively. The shared vision developed as part of the collaboration brief should be able to reduce the risk of conflicting expectations and goals. On the other hand, the agreed standards and procedures achieved during the planning of the solution should ensure the trustworthiness of information and effective communication and hence increase the level of trust and respect. Building mutual trust and respect in a collaborative venture should:

- enhance productivity by focusing attention on the work;
- encourage both individual and group interaction;
- minimise costs for monitoring;
- maximise individual and group energy and enthusiasm for the venture.

Conversely, a lack of trust and respect in a collaborative venture would:

- divert attention;
- stifle innovation;
- increase costs;
- drain energy.

In collaborative ventures some groups prefer to focus explicitly on building trust and respect as part of setting ground rules for collaboration whereas others allow these to evolve naturally through the group's interactions. Both approaches may be effective but have to be based on participants' preferences. Similar processes should be used here to those described in the 'bring the management team together' activity above.

Collate and disseminate procedures for collaboration

It is vitally important that all team members understand how they are expected to work and interact with each other during the lifecycle of the venture. A lot of effort has been spent so far to agree common processes and procedures, and failing to adequately disseminate these to the members can have a negative effect on the success of the collaboration. The collaboration champion and the members of the management team should be responsible for this task as well as ensuring that the procedures are adhered to.

A collaboration handbook should be developed, agreed upon and then disseminated to all members of the collaborative venture. This can be achieved by posting the document on the project extranet system as well as organising workshops and dissemination sessions as part of the project startup activities.

Provide management and technology training and mentoring as necessary

Training is seen by many as an important activity in the development of employees within an organisation. It is also equally important when developing, improving and maintaining interpersonal and intergroup collaboration. These are some of the key factors for the provision of training that relate to collaboration and other wider organisational issues:

- manage growth, expansion, modernisation and change related to, for example, the introduction of new technologies and working across organisations as part of collaborative projects;
- increase productivity and profitability, reduce cost and finally enhance skills and knowledge of the employee;
- prevent obsolescence;
- help in developing a problem-solving attitude;
- give people awareness of rules and procedures.

Training requirements for all participants in any collaboration should be carefully established by carrying out a skills audit against the prerequisites of any proposed technologies and systems to be used in the collaboration. It is vital that the cost and time needed in training participants to a level that will allow them to adequately use technologies for the collaboration is not underestimated. The training costs for many systems and ICT tools used in collaborative environments can form a substantial part of the overall cost of working collaboratively. In some circumstances training costs can be higher than the cost of purchasing software. To determine realistic training requirements, it may be helpful to talk to users and ICT managers from previous collaborations.

It may be necessary to maintain the levels of training through a continuous programme throughout the project's lifecycle; it is a common mistake to assume that skills gained in training sessions at the beginning of the project will be retained. This is also a good time to plan the ongoing training requirements for other corporate systems and for the rest of the employees in participating organisations.

Purchase and implement new technology

New technologies may be required for use by one or more members of the project team. These can often be implemented using upgrades or add-ons to existing hardware and software; most companies are reluctant to adopt new systems for single projects as the cost often outweighs the long-term benefit they would gain. The technology coordinator or systems manager should be responsible for the purchasing and implementation of any new systems for the venture.

When the project's ICT needs are identified, it is good practice to also establish the potential costs and long-term benefits associated with them. Decisions can then be made about how the cost of any new technology is divided among the team members. It does not necessarily mean that the participant who utilises a piece of technology should meet its associated costs, since its adoption usually enables the entire collaboration and benefits the entire project team. Information and communication technology costs and payment must be made very clear along with decisions about what happens to them after the project has ended. The implementation of any new system, be it hardware or software, will require physical implementation (installation) along with user training and mentoring.

Often, training in software systems can focus more on the way the software is used and less on the way the data within the system are managed across interfaces with other systems. In a collaborative environment it is critical that training covers all aspects of the collaboration requirements. Therefore, the information standards and procedures previously defined should form part of the ICT training strategy (see process 'provide technology training and mentoring as necessary' for more details on training for collaboration).

Reflections and feedback on collaboration

Learning lessons from each collaborative venture, whether they are positive or negative, is as important as any other aspect of collaboration. As working collaboratively often involves change in some processes, employees should be encouraged to freely express their views on their experiences throughout the collaboration venture.

The collaboration champion should conduct the post-venture review to determine and suggest areas for improvement in processes and procedures for the next collaborative venture.

The mechanisms for encouraging feedback must continue throughout the life of the project. Working practices and procedures for the collaboration may well change over time. Experiences and feedback should be captured through techniques such as observation, interviews or workshops on an ongoing basis. The model shown in Figure 3.7 is a typical method of conducting a review.

Collaboration support (from external sources)

It is often the case that, during the lifetime of the collaboration, a need for additional expertise arises within the venture or its associated organisations. When this occurs external support should be sought from specialist consultants. Typical reasons include:

- to supplement staff time;
- to supplement staff expertise;
- to ensure objectivity;
- to ensure credibility;
- to obtain a variety of skills;
- to deal with legal requirements.

A consultant benchmarking matrix shown in Figure 3.8 can be used to assess the skills of potential consultants based on required competencies and make decisions regarding their suitability for involvement in the project.

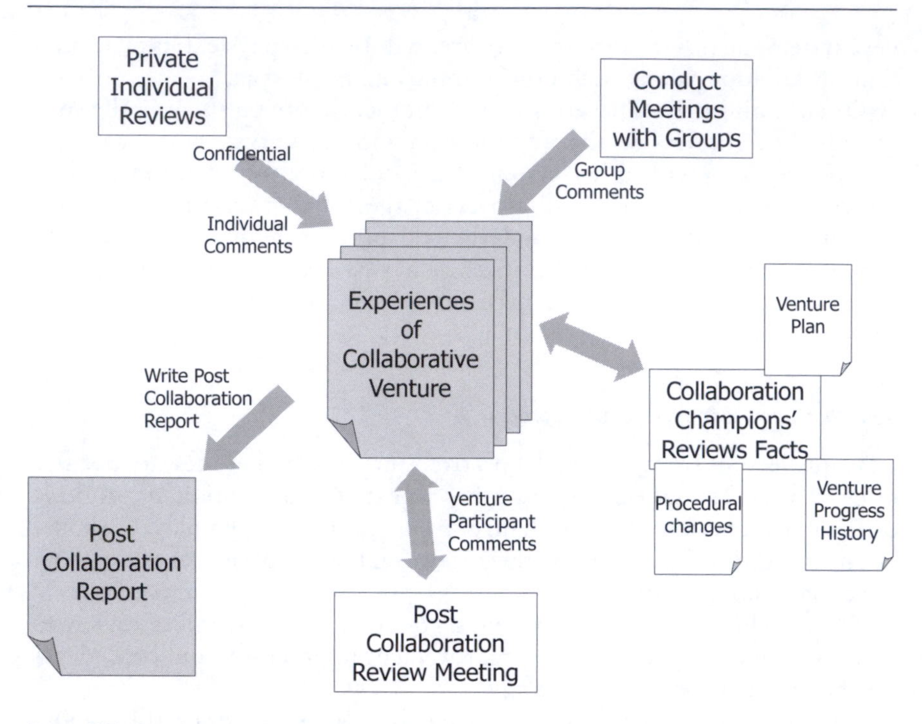

Figure 3.7 Conducting a post-collaborative venture review.

Measure collaboration performance

The aim of the monitoring process is to ensure that the objectives set out in the 'collaboration procedures' document remain on track, and that the success criteria agreed in the planning and implementing stage are met. A typical model used for measuring performance is shown in Figure 3.9.

COMPLETE PERIODIC PERFORMANCE EVALUATION

Evaluation reports should be prepared monthly by the collaboration champion. A standard method for reporting should be used. The reports should be discussed at monthly meetings with representatives from all collaboration partners.

INVESTIGATE SIGNIFICANT VARIANCES

Performance evaluation reports may show results that depart significantly from the standard set out in the 'set success criteria' stage. Where this is the case, the results should be examined and discussed and reasons for them sought. The reports may also indicate opportunities for improvement that should be followed up.

	experience of similar project (weighting total 20–30)				skills of key personnel (weighting total 15–25)					organisational resources and management systems (weighting total 30–40)			ability to support project financially (weighting total 20–30)							
	a	b	c	d	d	e	f	g	h	i	j	k	l	m	n	o	p	q		
Assessment Criteria	programme, method and approach	management and control procedures	resources to be applied to project	enviro, H&S factors	qualification and experience relevant to project	understanding of brief	flair, commitment and enthusiam	compatability with client and other team members	communication skills	organisation of team	authority levels of team members	planning and programming expertise	professional indemnity insurance	qa / management systems	workload and resources	ability to innovate	relevant experience	financial status	Total Weighted Factor	Consultant Ranking
competencies weighting																				
category weighting																				
Consultant																				
score																				
weighted score																				
score																				
weighted score																				
score																				
weighted score																				
score																				
weighted score																				
score																				
weighted score																				
score																				
weighted score																				

Figure 3.8 Consultants' benchmarking decision matrix.

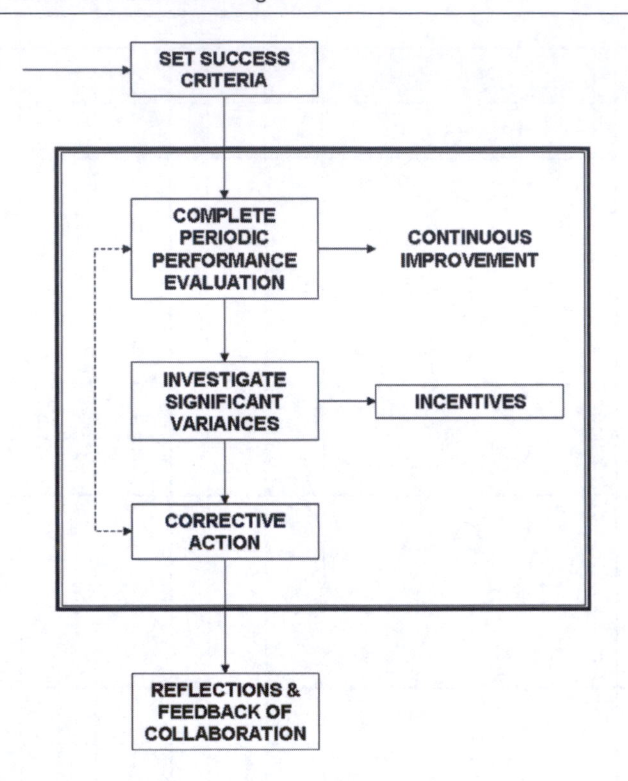

Figure 3.9 Performance measurement model.

INCENTIVES

Incentives should be considered for both organisations and individuals. In respect of individuals, performance reporting and issue identification should involve incentives. These are designed to stimulate a continuous interest and a measure of constructive competition. However, care must be taken to ensure that the benefits are fairly shared by all contributors. Significant savings in cost that result from improvement in performance should be shared in agreed proportions between the collaborating organisations. The arrangement should be included in the 'collaboration contract' documentation. This procedure should be based on calculations that can be audited.

CORRECTIVE ACTION

If the results are unacceptably below the norm, they have to be investigated and corrective action plans drawn up by the management responsible for the operation. These plans should be detailed and allocate specific responsibility. Table 3.6 provides a simple tool used to measure performance.

Table 3.6 An example of a performance evaluation form

	Score 0 (low) to 100 (high)
Alignment of purpose How well aligned are participants around the business context; overall visions and mission; goals and objectives; and priorities for the venture?	
Ability to perform How effective are participants at getting the job done; contributing their best; focusing on excellence; making a difference; ensuring individual and shared accountability for outcomes?	
Attention to process How effective are meeting management; progress monitoring and reporting; decision making; problem solving; conflict resolution; governance; and the internal/external measurement systems for your venture?	
Acuity of communication How well does the group exhibit openness/candour; use discipline and skill to provide and receive information; and ensure timely and accurate feedback in all aspects of the venture?	
Attitude of mutual trust and respect How well does the group share risk and reward; blend autonomy and interdependence; acknowledge and support each other	
Adaptability to learn and change How much attention is paid to continuous learning for all individuals; after-action briefings; institutional memory; continuous improvement?	
Total score (0–600)	

Key performance indicators (KPIs) and collaborative working

Tables 3.7 and 3.8 illustrate how the standard KPIs used for the construction industry and 'Respect for People' may be enhanced by collaborative working, the information in the right-hand column describing how the PIECC work can benefit each KPI.

Table 3.7 Industry KPIs

Client satisfaction – product	Improved processes will enable the client to receive an improved product
Client satisfaction – service	Improved planning and implementation will enhance the business process
Defects	Process defects may be eliminated
Safety	Communication is an important aspect of safety and risk management, and is a key output from PIECC
Construction cost	Improved planning and implementation will enhance the business process
Construction time	
Predictability – design cost	
Predictability – design time	
Predictability – construction cost	
Predictability – construction time	
Profitability	Elimination of waste will improve efficiency
Productivity	

Table 3.8 Respect for People KPIs

Employee satisfaction	Three of the four sub-measures – influence, achievement and respect – can be improved through application of PIECC
Staff turnover	–
Sickness absence	–
Safety	Communication is an important aspect of safety and risk management, and is a key output from PIECC
Working hours	More efficient planning may reduce time, or at least eliminate some of the waste
Travelling time	
Diversity	Collaborating with more organisations creates a diverse working environment
Training	Training is one of the key areas considered by PIECC
Pay	Improved planning and implementation improves employee performance
Investors in People	Training is one of the key areas considered by PIECC

Chapter 4

Information management and collaboration

Abdullahi Sheriff, Dino Bouchlaghem, Ashraf El-Hamalawi and Steven Yeomans

Introduction

Having established that developing and implementing an effective collaboration strategy is critical to improving project delivery and business operations, particularly in large non-collocated organisations and teams. This chapter presents an information management-based collaboration (IMC) methodology, a structured and practical approach for the design, selection, implementation and maintenance of a holistic strategy (including enterprise model, content model, systems and technology, implementation and change) to enable collaborative working within an organisation or within a specific construction project. It builds on the global vision for collaboration provided by the Planning and Implementation of Effective Collaboration in Construction (PIECC) framework by providing practical guides for developing focused information-driven collaboration strategies for construction projects or organisational processes.

This methodology consists of three frameworks: Framework A, Information Management Strategy; Framework B, Content Lifecycle Model; and Framework C, Metadata Standard (see Figure 4.1). Framework A provides a structured approach for developing a fit-for-purpose information management strategy for the whole project (and throughout its lifecycle). It begins by carefully defining the overall needs to ensure that there is a clear alignment between the holistic collaboration strategy developed and the operational project execution process (all the steps for this are explained in detail below). Framework B goes into further detail focusing on step 3 of Framework A. It provides a structured approach to defining the nature of information or content being created, managed and shared through the project lifecycle. Unlike Framework A, which focuses broadly on collaboration, Framework B focuses exclusively on the content model and the information systems. It is to be used during step 3 of Framework A. Framework C goes into further detail by focusing on the critical aspect of metadata to enable project stakeholders to develop a standard approach to naming, structuring and tagging all content generated through the project lifecycle. Thus Framework A provides the overall strategy for information management while Frameworks B and C complement it in task 3 by defining the nature of the content through the project lifecycle (Framework B) and then defining an appropriate metadata standard (Framework C). The interrelationship between all three frameworks is illustrated in Figure 4.1.

Executing the IMC methodology requires a good understanding of the overall project's needs and the participation of all stakeholders involved in delivering the project (including the client). If it is applied internally to improve a process within an organisation, care has to be taken to include the end users. Involving all stakeholders ensures that needs are correctly captured and understood, the strategy is well grounded and all parties involved or associated with the project buy into the collective vision for achieving effective collaboration. The three frameworks that constitute the IMC methodology are presented below.

Framework A: Information Management Strategy

Framework A, which consists of 12 critical tasks for defining an effective information management strategy for collaboration, is illustrated in Figure 4.2. This is executed in three phases representing solution definition, solution development and solution implementation. All the phases and each of the tasks are explained in detail below.

Phase 1: solution definition

The first phase focuses on defining the information needs of the project by understanding its nature and the overall delivery approach for which the collaboration strategy is required. This consists of the following three tasks.

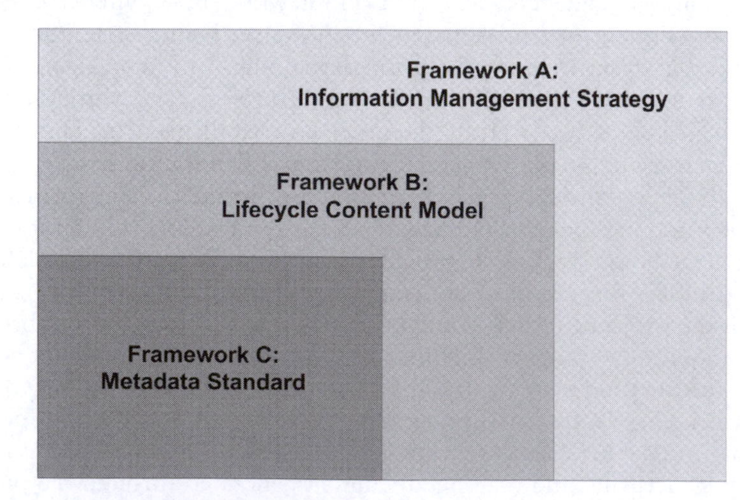

Figure 4.1 IMC methodology showing all three frameworks.

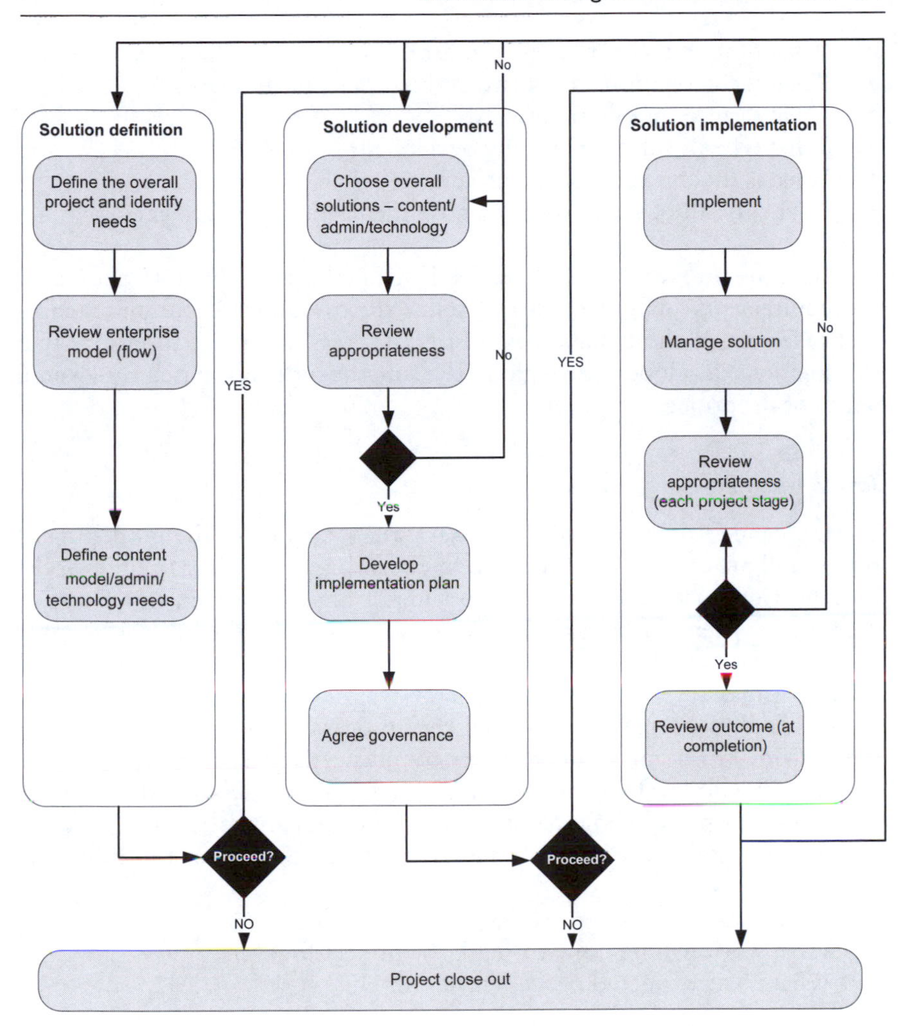

Figure 4.2 Framework A: Information Management Strategy.

Task 1: define the overall project and identify needs

The first task is to clearly understand the nature of the project being managed, its objectives, size, scope, stakeholders (including the client) and the services being offered. This sets the context for the overall project and enables the rest of the framework to be delivered in a manner that is effective, focused and fit for purpose. Various methods of data collection can be used for this, including workshops, questionnaires and interviews, but most importantly it should serve to answer the following:

- What is the overall objective of the project?
- Who are involved in the delivery of the project?

- What procurement process is in use?
- What is the relationship between the project participants?
- What services are being offered?
- What type of solution is being developed?
- What is the duration of the project?
- In light of all of the above, what are the needs of the project?

This is by no means an exhaustive list. Emphasis in this task should be on requirements analysis to clearly define the overall information management element of the collaboration brief. This is important to ensure that the solutions developed through all the subsequent tasks match the overall needs of the project.

Task 2: review the enterprise model

Having defined the needs of the project, the next task is the modelling of the overall process to be undergone in delivering the project in light of the findings from task 1. This is the most important task within this phase as it creates a visual diagram of the flow of work through the project lifecycle including the people involved, the various stages and the expected outputs. Whereas task 1 provides a general overview of the project, this task goes further by focusing on the delivery process itself to clearly map out the activities to be undertaken from commencement to completion. This mapping should be done to a relatively high level of detail, balancing the need to develop the whole picture with the complexity that such detailed mapping inevitably creates. The task is expected to answer the following basic questions:

- What work will be done through the project lifecycle?
- What is the expected process for doing this work?

It should be noted that the objective of the IMC methodology is not to redefine the enterprise model per se or to create a wholly unique project execution process. Instead, the emphasis is on ensuring that the information strategy aligns with the overall project delivery process. Therefore this task should ensure that the overall execution process and the individual activities are explicitly well defined, providing a firm basis upon which the content and technology needs of the project can be defined.

Task 3: define content model/admin/technology needs

The next step is to define the individual content, technology and administrative needs for each activity in the overall project delivery process as defined in task 2. It is important to focus on and establish the needs for each activity separately, as the needs of some activities may differ considerably from the needs of others. The objective here is to establish the ideal solutions required in light of the contextual constraints defined in task 1 and the work to be done defined in task 2. An example of how to implement this task is given in Framework B below.

The three tasks described above form the solution definition phase and serve to define the needs of the project and contextualise the type of solution required to achieve them. No decisions on the specific solutions to be used are required at this point. The next phase is the solution development phase, in which the specific solutions are developed and agreed.

Phase 2: solution development

The second phase, consisting of four tasks, is when the specific decisions are made regarding the solutions to be used to deliver an effective information management strategy for the project based on the information needs defined in phase 1.

Task 4: choose overall solutions

Having defined the content, admin and technological needs for each stage of the project, the fourth task involves rationalising these options and choosing the appropriate solutions. It is important to commence with an appreciation of the overall context and needs of the project prior to selecting the solution to be used as it enables informed solutions to be identified. The needs defined in task 3 will often give rise to a diverse and conflicting range of possible solutions. Thus this task is carried out with a view to selecting the optimum solutions for the project to suit the needs and the specific context of the project (time frame, budget etc.). This may result in either a single set of solutions or a range of solutions suited to each stage of the project.

For example, it could be established that, based on the needs defined for all the activities to be carried out, only one type of content should be used throughout the project: text files, created in Microsoft Word and submitted in portable document format (PDF). Alternatively, it could be decided that, as only a single organisation is involved in delivering the first three stages of the project, no web-based information system is required for storing and managing information as the organisation's existing internal file sharing with an appropriate folder structure is sufficient. From Stage D however, a large number of participants will be introduced into the project, therefore a web-based content management system will be required capable of managing PDF (the type of content to be used through those stages).

Irrespective of the decisions made, grounding such choices in the right operational context of the project will ensure that the right type of solutions are found and implemented to support the project. The selections should answer the questions:

- What content is required to deliver the project?
- What systems are required to create, manage, share and store these?

Task 5: review appropriateness

Having chosen the possible solutions, the next task is to review their appropriateness. Here, the chosen type(s) of content, systems/technology (or suite of technologies) for managing the content, and administrative infrastructure are reassessed in light of the enterprise model, the project needs that have been defined and the possible scenarios to be undertaken in executing the project. This validation is essential to ensure that the solutions are indeed the right ones. Where the solutions are deemed to be appropriate and signed off by the project leadership, effective implementation plans should then be developed (task 6). Solutions are sometimes deemed inappropriate because the project needs are poorly defined, in which case phase 1 is repeated, rethinking the rationale for collaboration and the overall objectives of the project. Alternatively, the overall outputs from phase 1 may be well defined but the choices made in task 4 may be inappropriate. In such a case, task 4 is revisited to reconsider the choices made and select a more appropriate solution. Irrespective of the method used to conduct this validation, it is critical to ensure that the criteria used for assessing the appropriateness are based on the needs defined in phase 1, thus aligning the solutions developed to the overall needs of the project.

Task 6: develop the implementation plan

The next step is to develop the method for effectively delivering the collaboration solutions, implementing it effectively in the project and managing the change into the new ways of working. In doing this, it is essential to understand the cultural, social and political dynamics of the stakeholders working together to deliver the project, the client and the project location. This should also be developed to take into account the type of solutions to be adopted and what is needed to move from concept to full use. The plan should focus on the technical (standards, systems, software) and functional/operational (procedures, administration, ownership) aspects as well as people (training, support, incentives). No specific actions should be carried out at this stage. Only a plan of how the implementation will be carried out is required.

Task 7: agree governance

To support the implementation plan and ensure that the solution developed is sustained, the next task is to agree the overall approach to managing and owning the solutions throughout its expected lifecycle. This task is essential as governance creates ownership and clearly assigns responsibilities. It also ensures that the cost, time and operational impact and implementation plan of the proposed solution are clearly discussed and appreciated by all the stakeholders prior to implementation. Governance should include the management of:

- content and ensuring that the content solutions defined are utilised effectively and maintained;
- all procedures and processes for administering the collaboration strategy through the project lifecycle; and
- systems and technology implemented to improve collaboration.

The governance approach should also include explicit responsibility for effective change management, to ensure that there is a smooth transition into the long-term use of the solutions. It should also include the measurement criteria to be used in periodically assessing the performance of the IMC strategy through the project lifecycle. A framework for change management is given in Chapter 8 of this book.

The above four tasks form the solution development phase, in which the specific solutions to enable information management for collaboration on a project are outlined. On completing this phase, the solution developed, the implementation plan and the governance should be reviewed to assess if they collectively meet the overall project needs (as defined in phase 1) and therefore agree if the use of the methodology should proceed. If the solution is deemed appropriate, the project then moves on to the final phase: solution implementation.

Phase 3: solution implementation

This phase of the process is concerned with the practical implementation, management and review of the solution and includes four tasks.

Task 8: implement solution

The solutions developed in the previous phase should be implemented in line with the implementation plan developed in task 6. This implementation process will vary from one project to the next and will depend on the solutions selected, the people involved in the project, the type of project, the technical complexities of the system used and so on. Also carried out here is the actual procurement and set-up of the systems to be used (if any) to manage the content through the project lifecycle. The training plan also defined in task 6 is executed, ensuring that all end users and stakeholders become fully aware and au fait with the collaboration strategy. It is important to pay particular attention to user needs at this point and to ensure that change is managed effectively, as only a solution that is well accepted and understood by all participants is likely to be successfully implemented. When the solutions are modified on account of changing project needs, new training is also required to ensure that users are fully aware of the new strategy and to sustain their buy-in.

Task 9: manage the solution

As such projects will often run over a period of time, this task effectively manages the solutions developed and implemented to ensure that they continuously remain functional and effective in line with the overall needs of the project. This includes the administrative maintenance of any systems implemented to enable collaboration as well as the regular training and retraining of people to ensure compliance. The governance structure agreed in task 7 serves as the basis for managing the solution.

Task 10: review appropriateness

At each stage of the project, the information strategy should be reviewed in light of the defined project needs and new variables that may develop throughout the project lifecycle to ensure that it remains the most appropriate solution for the next phase and for the overall project. This task therefore serves to complement task 9. For example (in a project being executed based on the Royal Institute of British Architects plan of work), the briefing stage of a project may require a different information collaboration strategy for the conceptual design stage, which may in turn be different from the one required for the detailed design stage and so on. Therefore, the overall strategy should be reviewed when moving from one stage to the next. Although the project needs from all the stages should have been defined in tasks 1 and 2, changes do occur during the project process which may require a rethink of the solutions in use. This task therefore ensures that, at the end of each such project stage or phase, the solution is revisited in light of the needs of the next phase and the overall project to ensure that it remains the most appropriate to effectively deliver the project. A solution may become inappropriate because of changes in the needs of the project; thus tasks 1 and 2 should be repeated to redefine the overall project needs. If no changes emerge in the project needs but more suitable systems or changes in the governance/implementation model are identified to improve the effectiveness of the collaboration strategy, then only phase 2 (tasks 4, 5, 6 and 7) is repeated. The objective is to ensure that only the most appropriate solutions are used for delivering the information and collaboration strategy. If, however, the solutions are appropriate and no changes are required, the project proceeds as planned until the next phase or the project is completed.

Task 11: review outcome

At the completion of the project, it is important to review the overall process to learn lessons, identify weaknesses and reinforce strengths on future projects. This will ensure that future collaboration strategies are better designed and optimised to improve the project's effectiveness. A number of methods may be used to achieve this, including workshops, focus groups and questionnaires. Each organisation or project should use the approach most suited to its operations, culture, needs and circumstances. What is important is to ensure that all lessons identified are acted upon.

Task 12: project close

This is the formal end of the project and typically would occur following task 11. However, the project may be closed out much earlier if the output from the project definition stage shows that no collaboration solution is required or if the needs of the project are not effectively defined. Similarly, the solution developed in phase 2 may result in a decision not to proceed for various reasons including lack of effective governance or an inability to deliver the required solution. It is important to assess these needs at the end of each phase. The project leadership should make the decision whether or not to proceed in line with the information obtained and the specific context of the project. Where the methodology is employed in full and all 11 tasks are completed, the process used to develop an information collaboration strategy should be declared complete only after the lessons learnt have been captured and sufficient steps have been agreed to assimilate all such lessons back into the overall project process (i.e. task 11).

Framework B: Content Lifecycle Model

This framework is illustrated in Figure 4.3 and provides a process that can be used to complete task 3 of Framework A by focusing on the typical flow of information for an activity.

For each activity and at each task within Framework B, answers are sought to the following questions:

- What type of content is used?
- How will it be administered and who will be responsible for it?
- What type of system is required for sharing it?
- What type of system is required for storing it?

As an example, suppose that the process analysis conducted in task 2 highlights 'conceptual design' as an activity. Using Framework B, this activity is then analysed in detail using the questions above to identify the type

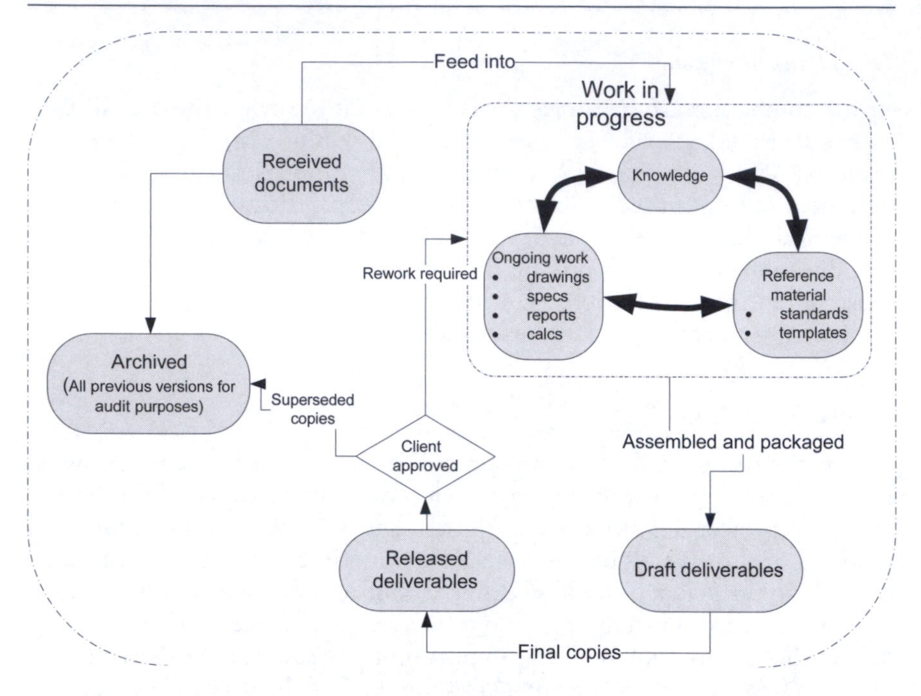

Figure 4.3 Framework B: Content Lifecycle Model.

of information received, created, shared and delivered to the client while carrying out this activity. What is required in this step is the definition of the 'type of' content or technology required. For example, on a residential design project, it suffices to say that two-dimensional computer-aided design drawings will be the most common type of content, produced and shared electronically. No further details are required on its specific format.

Framework C: Metadata Standard

Following the definition of the content model which clearly explains the nature of content on the project, Framework C is used to define the metadata standard to be used to provide consistency in tagging and structuring content throughout the project lifecycle. Framework C is illustrated in Figure 4.4 and consists of 12 tasks grouped under the three phases: project definition (tasks 1–4), metadata development (tasks 5–8) and implementation (tasks 9–12).

Phase 1: project definition

This phase aims to contextualise and position the metadata development project, ensuring that subsequent phases are both precise and adequate to meet the defined goals. It includes the following four tasks.

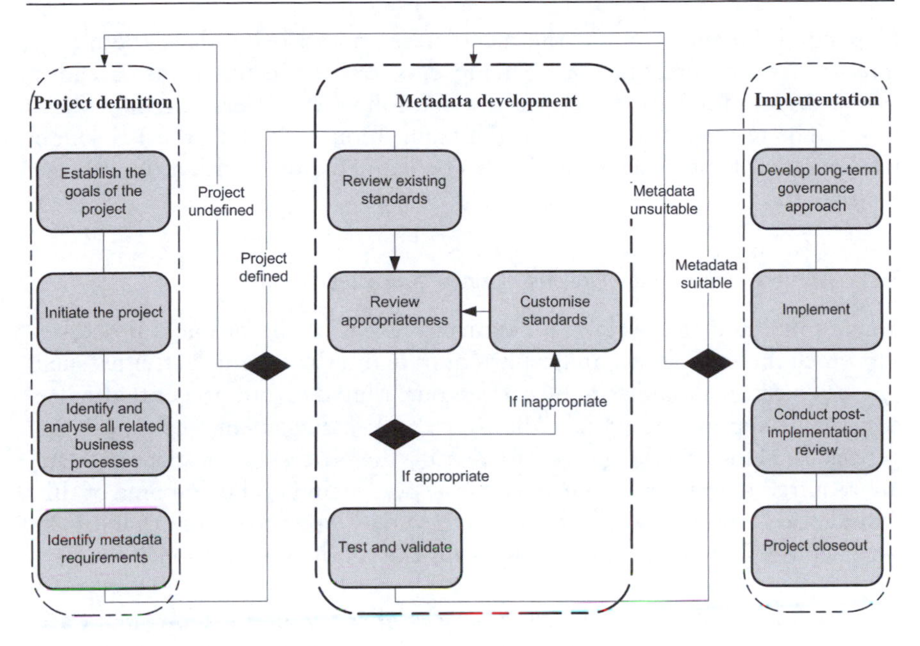

Figure 4.4 Stages for developing an organisational metadata standard.

Task 1: establish the goal of the project

The first task is to clearly define the intended goals of the metadata development project to ensure that appropriate actions are taken and resources are made available. It also enables a strategic needs analysis to be carried out eliciting specific answers to questions such as:

- Why should this project be carried out?
- What is the desired outcome?
- What is the scope of the project?
- What business streams across the company will be affected by the project?
- What specific content classes are intended to be managed?
- Who are the target end users?

Task 2: initiate the project

The overall objectives should be translated into an actionable plan to articulate how the metadata development project will be managed from inception to completion. Here, the business case and communication plan should be developed including detailed justifications of the viability and cost of the project. A high-level champion will need to be appointed to provide senior management support. A project delivery team will also need to be set up with a clearly defined mandate and responsibilities. In appointing the team (depending on the scope of the metadata standard being developed) it is important to ensure that membership cuts across the functional breadth of the organisation to reflect the distribution of end users and gain their input. The resources available to the team (monetary and other) through the lifecycle of the project should also be defined, along with a time frame for execution.

Task 3: identify and analyse all related business processes

Having initiated the project, a detailed analysis of the business processes for which the metadata standard is required and the specific activities such processes entail should then be carried out. Metadata aim to contextualise content by supporting its use, discovery and management to support the processes. Hence this stage aims to identify the processes that the metadata are required to support (from an end user perspective). The outcome of this would also be used to validate the standard developed to ensure that it is fit for purpose. This stage seeks answers for the following questions:

- What are the specific business processes for which the metadata are required?
- Within each of those processes, what activities are carried out which require metadata?
- What criteria should be used to determine what standards are appropriate for the company's needs?

Task 4: identify metadata requirements

Based on the understanding of the processes, the nature of the metadata required should then be identified. It is important to maintain the order of the tasks, starting with an understanding of the processes and then building up a picture of the sort of metadata needed to support it. At this stage no details are required on the individual attributes. Instead themes such as subject metadata, administrative metadata, regulatory metadata, retrieval metadata and workflow metadata are required.

Phase 2: metadata development

This phase translates the project vision into a usable/actionable standard. If the standard developed is deemed suitable for the company (based on the needs defined in phase 1) the project then proceeds to the final phase. If, however, the attributes are deemed inappropriate for the defined company needs, either more suitable attributes need to be defined (by repeating phase 2) or the company needs and goals for the project need to be reassessed (by repeating phase 1). This phase consists of the following four tasks.

Task 5: review existing standards

Standards should be reviewed to identify which (if any) can be adopted to meet the needs of the company and its business processes (as defined in phase 1). As broad as possible a range of standards should be consulted, based on their international applicability, relevance to the organisation, relevance to the industry in which the organisation is based and/or relevance to the type of content being managed. Whereas these selection criteria may vary across companies and indeed projects, they need to be explicitly defined. Some key questions which should be answered here include:

- What international/cross-industry standards are available which meet the defined criteria (from phase 1)?
- What industry-specific standards are available which meet the defined criteria?
- What content-specific standards are available which meet the defined criteria?

Task 6: review appropriateness

The individual attributes contained within each of the standards shortlisted should then be analysed to determine their suitability for the specific needs defined in phase 1. Where a standard is deemed fully appropriate and can be adopted with little or no customisation it can be carried forward to the testing stage. Where customisation is deemed necessary, attributes which can be adopted as a base minimum should be identified first. These then serve as a good starting point from which the required customisation can be carried out. The metadata attributes adopted are also refined here, identifying variations such as compulsory and optional attributes or automatic and manual attributes. The review process should be carried out iteratively until the developed standard is deemed appropriate.

Task 7: customise standards

Where no existing standards are fully appropriate, the baseline identified in task 6 should then be customised to create a bespoke standard which reflects the company's needs. This should begin with a clear understanding of why the customisation is necessary. If carried out appropriately, this will result in a company-specific metadata standard which takes into account different aspects such as business processes, archiving policies, quality management procedures and business structure. Each attribute identified here must be justified, highlighting why it is necessary and for whom, to ensure that the standard developed remains fit for purpose. Various methods can be used to conduct this, including desk studies of existing repositories, workshops and questionnaires. Irrespective of the method employed, the eventual outcome must be collectively reviewed by the project team, accepted and signed off prior to any testing or validation.

Task 8: test and validate

To ensure its suitability, the metadata standard should then be tested using various scenarios. These should be as varied as possible but should reflect the expected use cases for the completed standard. The object of this exercise is to scrutinise the standard for any loopholes and ensure that the solution is robust enough to meet the needs of the company and its wider user community.

Phase 3: implementation

This phase includes the practical set-up, management and evaluation of the standard and consists of four tasks.

Task 9: develop governance approach

Governance ensures accountability and responsibility for the long-term management of the metadata standard. This is necessary in metadata development and implementation to ensure that it is continually updated to meet the future needs of the company and its users regarding the evolving business strategies, working methods, processes and regulations. It requires specific decisions to be made, including (but not exclusively) how the standard will be managed in the future; who retains responsibility; how the quality of metadata input into the system can be ensured; and what resources will be committed for this purpose.

Task 10: implement

The developed standard is encoded into the organisation's software systems to make it machine actionable. Also considered here are the visual interfaces; visualisation of the metadata in the system (as seen by the user); the interfaces for metadata entry; result visualisation; and automating attributes. Beyond the technological implementation, training and policy guidelines developed earlier are also implemented here along with effective change management, all of which are required to facilitate a smooth transition to the use of the new standard.

Task 11: conduct a post-implementation review

The processes undergone and the outcome of the project should then be collectively reviewed to ascertain if the original project goals have been met. The reflection process also enables the lessons learnt to be recorded and disseminated, thus improving the delivery of future metadata development endeavours. Feedback should be regularly obtained from end users to establish the state of use and the appropriateness of the metadata standard for their ongoing needs.

Task 12: formal close

Finally, the formal close of the project concludes the process.

Conclusion

The three frameworks presented above make up the Information Management and Collaboration methodology, aimed at developing effective information management strategies for enabling collaborative working in an organisation or on a project. It reinforces the need to align the solutions to be used to the organisation's overall needs and provides rigour in achieving this. It also emphasises the need to critically analyse processes to understand underlying information needs, thus grounding the choice of solutions on stakeholder requirements. Using such an approach will create a strategy which is effective, fit for purpose and supportive of the overall project delivery process.

Chapter 5

Mobile communications and wireless technologies

Ozan Koseoglu and Dino Bouchlaghem

Introduction

Mobile computing and wireless networking have experienced major developments and advances over the last decade. Many industrial sectors now use mobile technologies and wireless links as part of their daily business. The nature of work is constantly changing as professionals spend less time at their desk and require mobile or flexible working models and technologies for communication and collaborative working. Mobile technologies are becoming key enablers for collaborative decision making supported with the instant and remote access to data. These technologies are enabling the concept of mobility in business, giving communication service providers new application opportunities in which, for example, data and voice can converge through wireless networks.

Wireless networks

A wireless local area network (WLAN) is a flexible communication medium that serves as an alternative to a wired local area network (LAN) and uses wireless technology to connect computers and servers. WLANs transmit and receive data over the air space using electromagnetic waves, with the basic components being the wireless station, access points, antennas and adaptors. Every computer or device that transmits or receives data over the wireless network is called a station. An access point is a device whose purpose is to receive radio transmissions from other stations on the WLAN and send them to the wired network. Antennas carry radio signals and extend the coverage area created by the access points. Users access the WLAN through their wireless LAN adaptors. These adaptors can be either inserted as PC cards in notebooks or PCI adaptors in desktop computers, or included as integrated devices within handheld computers (Wireless LAN Alliance, n.d.). There are two types of WLAN configurations:

1 *independent WLAN configuration*, in which a set of personal computers equipped with wireless adaptors and within a certain range are

connected to set up an independent network (Figure 5.1; Bruce and Gilster, 2002; Wireless LAN Alliance, n.d.);

2 *infrastructure WLAN configuration*, in which an access point links the wireless to the wired network, enabling users to share resources (Figure 5.2). Here, the type of access point determines the coverage area for the wireless network, which can be in a private home, an office floor, a building or a campus (Wireless LAN Alliance, n.d.).

WLAN design

The design of a WLAN is important when implementing a WLAN within an existing corporate network. The needs of a company need to be carefully established to identify a design that suits both the budget and business processes. A site survey and an initial selection of equipment required should form the preliminary steps. The site survey is conducted to define the coverage area and network size and establish the network capacity, radio frequency and power requirements.

The signal coverage should be tested in order to determine the network connectivity and the number of access points. It is important to bear in mind that the signal coverage can be affected by various factors such as linear distance from the access point (AP), the power provided by the AP,

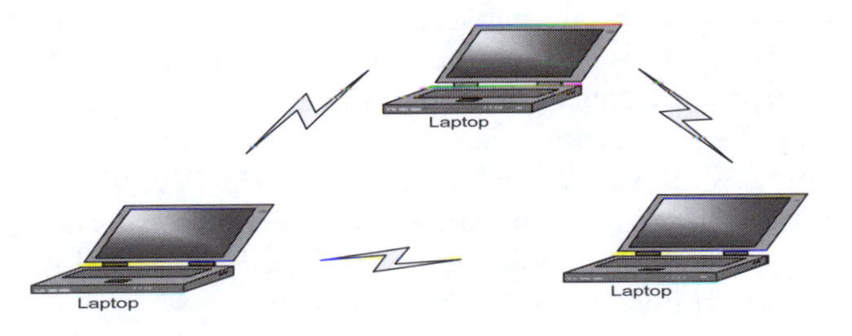

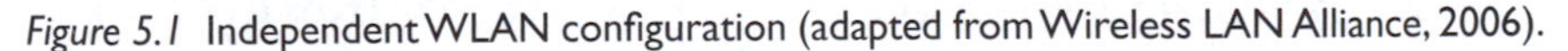

Figure 5.1 Independent WLAN configuration (adapted from Wireless LAN Alliance, 2006).

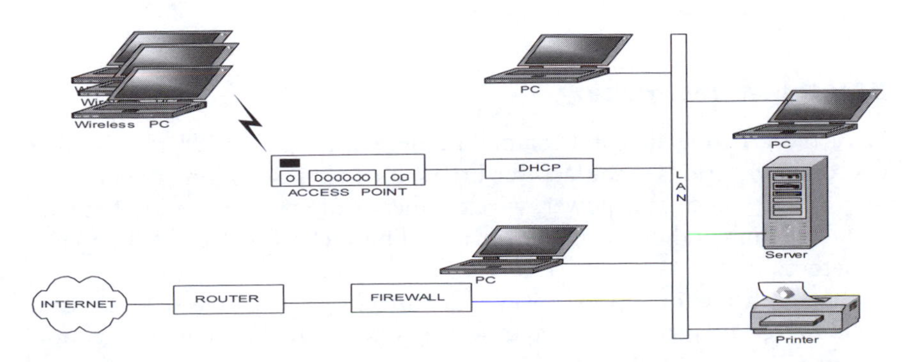

Figure 5.2 Infrastructure WLAN configuration (adapted from Bruce and Gilster, 2002).

the antennas used on the access points and the stations, the wall/floor/ceiling materials of the enclosure, the shape of the building, the network area, the number of rooms in the building and any physical objects within the internal space.

The network capacity should take into consideration the number of users, the organisational structure, the software used, the expected company growth and the type of network services. The total cost of equipment, capacity, security, technical support, installation, setting up and maintenance has to be determined before selecting and installing the network components.

WLAN security

A number of techniques are available to secure wireless networks: the most commonly used are Media Access Control (MAC) address filtering and Wired Equivalent Privacy (WEP). Access points can be configured for MAC filtering, so that only users with MAC addresses on a predefined list are allowed into the network. The WEP algorithm provides protection against eavesdropping by encrypting the data using a 128-bit key (Intel, 2003).

These protocols may not be applicable to every condition; the list below gives some possible precautions that can be applied:

- placing internal firewalls between the wired LAN and WLAN;
- changing passwords and Internet Protocol addresses;
- preventing radiofrequency leakage outside the area;
- changing WEP encryption keys;
- combining WEP with virtual private networks (VPNs) for more effective protection (a virtual private network enables users on an insecure network, such as the internet or a WEP-based 802.11 WLAN, to establish a secure connection with the network);
- establishing an authentication dial-in user service (RADIUS) server on the wireless network; the RADIUS server controls and verifies the validity of the users before entering the network.

Bluetooth networking

With the growing demand for mobile networking, in 1994 the Swedish telecommunications manufacturer Ericsson began researching the possibilities of a low-cost, low-power, wireless means of communication between mobile phones and other mobile devices. This technology is now known as 'Bluetooth'.

Bluetooth technology is a way of connecting electronic devices using radio waves to provide fast and secure voice and data communication.

The maximum range for connection is 10 metres and a line of sight is not required. Bluetooth technology can provide both point-to-point and point-to-multipoint connections. In the point-to-point connection, communication is established between the two devices, whereas in the point-to-multipoint connection the communication channel is shared by up to eight units at most. Bluetooth uses base-band transmissions, which convert digital and voice data into formats that can be transmitted using a radio signal. The major benefits that Bluetooth provide include a low-cost wireless network solution, low power consumption, global compatibility and instantaneous and transparent connections. Bluetooth products can be added to an existing device in order to enable Bluetooth connectivity. These add-on products include PC cards, and USB/memory/phone adaptors. However, the Bluetooth technology has some disadvantages such as the short range (up to 10 metres) and the limited data transmission speed (1 Mbps).

Third-generation technology (3G)

Over the last few years, phone companies and telecom operators have invested heavily to acquire the third-generation wireless technology (3G) licences from national governments. These licences enable them to provide wireless services within a specific frequency band and in a certain geographical area. However, some of these companies went bankrupt while trying to make huge payments. Other companies still in business could not afford to develop their third-generation networks because of the economic slowdown in the telecommunication sector. Alternatively, some operators developed 2.5-generation technology (2.5G) as a cheaper solution (*PC Magazine*, 2001).

The 2.5G technologies were built on the existing second-generation (2G) GSM networks and have some of the benefits that will be used in 3G such as faster data transfer, provision of internet connection and continuity of service. High Speed Circuit Switched Data (HSCD), General Packet Radio Service (GPRS), Enhanced Data Rates for Global Evolution (EDGE) and IS-95B are all 2.5G technologies. They were all designed to work on GSM networks owing to its popularity all around the world as a 2G standard. The most used 2.5G technology around the world is GPRS, which can achieve a maximum transmission speed of 171.2 kilobits per second (kbps) (Andersson, 2001; Steward *et al.*, 2002). GPRS can support different types of visual information and still images including photos, postcards, traffic location, maps and graphs.

Third-generation wireless technology includes 3G, 2.5G and 2G functionalities. Video on demand, videoconferencing, high-speed multimedia and mobile internet access are all important features of 3G. Data transmission rates are much higher than what can be achieved in existing networks

(GSM Association, n.d.; Land Mobile, 2005). The 2G systems in use today can transmit data at around 10 kbps. The 2.5G system's rate ranges from 28.8 kbps to 144 kbps. 3G systems in the future will be faster, with a speed ranging from 384 kbps to 2 megabits per second (Mbps). This worldwide wireless connection is compatible with the existing mobile telecommunication networks. New devices for 3G communications have already reached the market. These devices include mobile phone capabilities and provide interactive videoconferencing (Intel, 2004).

Third-generation-enabled mobile phone systems were introduced in the year 2001/2. The aim for 3G was to unify the 2G wireless networks in order to develop a common standard. However, the reasons for the uncertainty about 3G concepts and development are related to the following:

- The form of mobile communications has radically changed, extending to non-voice applications.
- 3G licences have been awarded by many governments around the world at huge costs.
- 3G is based on a different technology, which is called Code Division Multiple Access (CDMA), whereas 2G is based on Time Division Multiple Access (TDMA).
- Many industry experts have questioned the return on investment of 3G technology.

Another difficulty with 3G networks is billing, because their structure is fundamentally different from existing architecture. To create an efficient 3G billing system, operators must take both the following factors into account (*PC Magazine*, 2001):

- 3G uses packet-switched technology, which means that users are always connected and online.
- The future expansion of the services and capacity to prevent overcrowding in today's 2G networks.

Worldwide interoperability for microwave access

Radiofrequency-based broadband access networks are rapidly developing in wireless and mobile communications because of the emergence of multimedia applications, visualisation and demand for high-speed internet access. Therefore, new wireless networks with broadband capabilities are being sought to provide high-speed integrated services (data, voice and video) with cost-effective support for quality (Khun-Jush *et al.*, 2000; Intel and WiMAX, 2004). The term 'WiMAX' (worldwide interoperability for microwave access) has become synonymous with the IEEE 802.16 wireless metropolitan area network (MAN) air interface standard. WiMAX fills the gap between wireless LANs and wide area networks (WiMAX Forum, 2004).

Wireless broadband access is set up as cellular systems using base stations which provide a service in a radius of several miles/kilometres called 'cells'. Base stations do not have to reside on a purpose-built tower. They can simply be located on the roof of a tall building or another high structure such as a grain silo or water tower. The original 802.16 standard operates in the 10–66 GHz frequency band and requires line-of-sight towers.

In January 2003, the Institute of Electrical and Electronics Engineers (IEEE) approved the 802.16a standard, which covers frequency bands between 2 and 11 GHz. This standard is an extension of the IEEE 802.16 standard for 10–66 GHz published in April 2002. These sub-11 GHz frequency ranges enable non-line-of-sight performance, making the IEEE 802.16a standard the most appropriate technology for last-mile applications (Intel and WiMAX, 2003).

The most common 802.16a configuration consists of a base station constructed on a building or tower that communicates on a point-to-multipoint basis with end user stations located in businesses and homes. 802.16a offers a range of up to 30 miles with a cell radius of 4–6 miles (Intel and WiMAX, 2003).

WiMAX covers wider metropolitan or rural areas whereas WLAN (802.11a, b and g) handles smaller areas, such as offices or hotspots. It can provide data rates of up to 75 Mbps per base station with typical cell sizes of 2–10 kilometres (Intel and WiMAX, 2005). It is an emerging technology that will deliver broadband connectivity in a larger geographic area than WLAN and enable greater mobility for high-speed data applications (Intel, 2004). Therefore this is filling the gap between WLANs and wide area networks such as GSM and 3G (WiMAX Forum, 2004).

Wireless protocols

Wi-Fi (WLAN), WiMAX, 3G and Bluetooth technologies are all necessary to form the global wireless infrastructure needed to deliver high-speed communications and internet access worldwide. WLAN is ideal for isolated connectivity whereas WiMAX and 3G are needed for long-distance wireless communications (Figure 5.3). 3G services will enable highly mobile users with laptops and other wireless data devices to bridge the gap between higher bandwidth WiMAX hot zones and Wi-Fi hotspots (Intel, 2004). Bluetooth is the leading technology for short-distance instantaneous communications between mobile devices (Figure 5.3). It best suits end users who use more than one mobile device. It provides data and voice communication without any specific installations required.

Wireless protocols are tabulated in Table 5.1 according to their data rate, coverage area, system requirements and installation cost items.

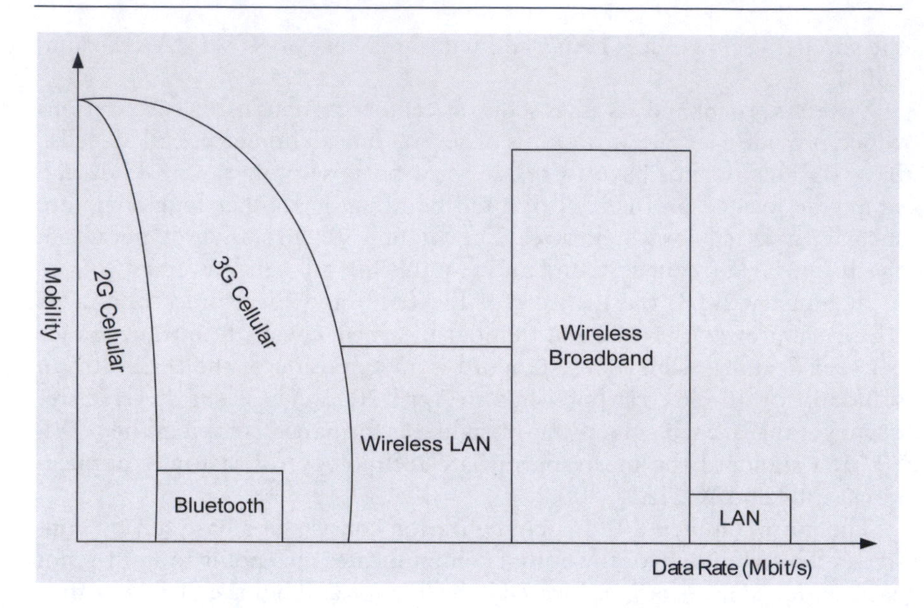

Figure 5.3 Mobility versus data rate graph of wireless technologies.

Table 5.1 Comparison of wireless protocols

Mobile technology	*Bluetooth*	*WLAN (802.11g)*	*3G (UMTS)*	*WIMAX (802.16a)*
Data rate	1 Mbps	54 Mbps	384 kbps	75 Mbps
Coverage	0–10 m	Limited with access points	Extensive coverage	0–50 km
System requirements	Bluetooth equipment	WLAN equipment	Telecom operator	Service provider
Cost	Hardware components	Hardware components	Hardware and service fee	Hardware and service fee

Voice over Internet Protocol

Voice over Internet Protocol (VoIP) enables businesses to make phone calls across computer networks and provides a low-cost and efficient way to complement traditional phone systems. VoIP can be used in individual local office networks or between sites, enabling the integration of call handling and reporting with other business processes and with websites. VoIP can also deliver telephone services to remote end users and home workers by using a broadband connection to the internet (Figure 5.4) (DTI, n.d.).

Traditional phone calls use the entire phone line for each call whereas in VoIP voice data are compressed and transmitted over a computer network. This means that VoIP uses up to 90 per cent less bandwidth than a

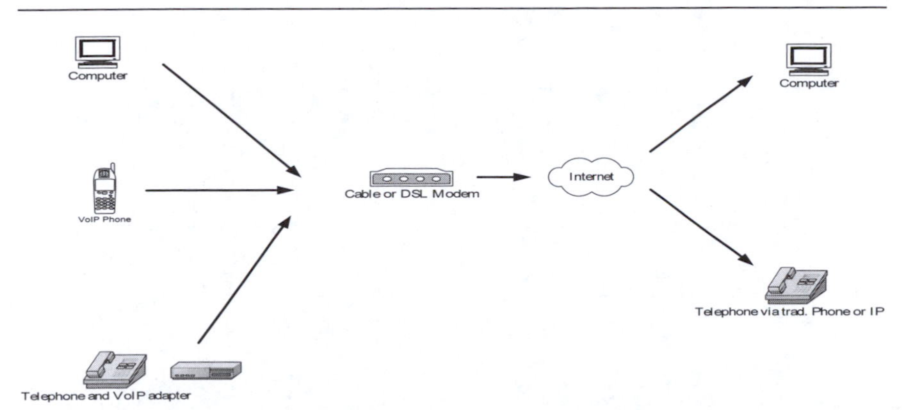

Figure 5.4 VoIP structure (iDesk, n.d.).

traditional phone call and is more cost-effective and more efficient (DTI, n.d.). The major components of a VoIP network are the call-processing server (IP PBX), end user devices, media/VoIP gateways and the IP network (Brunner and Ali, 2004).

The call-processing server, known as IP PBX, is the heart of a VoIP phone system and manages all VoIP connections. The end user devices consist of VoIP phones and desktop devices. VoIP phones may be software-based 'soft-phones' or hardware-based 'hard phones' such as traditional phones. The major function of the media gateway is the conversion of voice from analogue to digital and the creation of voice IP packets (Brunner and Ali, 2004). The immediate benefits of VoIP for business can be cheaper external calls, free internal calls, simple infrastructure and improvement in flexible working (DTI, n.d.).

The fast growth of the two leading technologies, wireless LAN and VoIP, has come together to provide a new application, VoIP over wireless LAN (VoWiFi). VoIP over wireless LAN enables seamless communications virtually anywhere in wireless-enabled environments, public hotspots and home offices (Intel, 2005; Netlink Wireless Telephone Portfolio, n.d.; Poulbere, 2005) (Figure 5.5). VoWiFi will offer cheap voice services to mobile subscribers located in areas covered by Wi-Fi. Another driving force for the development of VoWiFi is the availability of mobile handsets integrating Wi-Fi. By using the VoWiFi system, end users can:

- make voice calls over the broadband connection in their home;
- connect wirelessly to the office phone system;
- connect to the company telephone network from a public hotspot;
- connect to the office phone network when visiting other company locations; and
- make low-cost calls from public hotspots in cafes, airport lounges and the like.

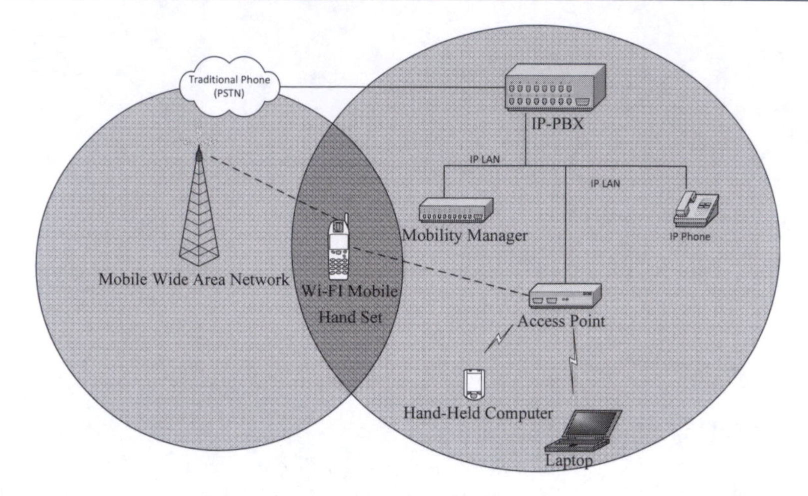

Figure 5.5 Integrated infrastructure of Voice over Wi-Fi and existing networks (Proxim Wireless Networks, 2004).

Mobile end user devices

Mobile devices which are used in wireless networks are developing rapidly, getting smaller and more sophisticated. The latest end user terminals used for wireless networking are mobile or smart phones, laptops, PDAs (personal digital assistants), Pocket PCs and tablet PCs.

Mobile phones for 3G communications are available and provide features such as interactive video conferencing, PDA capabilities and integrated wireless protocols such as 802.11b and 802.11g (Intel, 2004). Devices that combine the features of mobile phones and PDAs are becoming popular as they enable users to access email and the internet, conduct videoconferences, view documents, transfer data and make calls on the same device (Computer Associates International, 2002).

A PDA is a handheld digital organiser that stores contact information and allows users to take notes. PDAs are basically used for information management. They can run versions of office software to open and edit documents, spreadsheets or other files from the office. A Pocket PC has many types of wireless device options. The properties of the Pocket PC are similar to those of the notebook computer. They have large full-colour screens and larger memories. Some of the PDA manufacturers develop their own programs, but unlike Palm models, which use PalmOS, Pocket PCs use a version of Pocket Windows as the main software. The latest development in mobile computing is the tablet PC, which is an ultra-compact lightweight laptop which can be held like a book and can be used while seated or standing. New models of PDAs, Pocket PCs and tablet PCs are now available with integrated Bluetooth and WLAN-enabled options (Compaq Official website, 2004).

Wireless and mobile technologies in construction

Mobile computing and wireless networking are emerging technologies that are currently helping construction companies improve distributed and collaborative working in various areas of their business activities. Construction site processes, for example, can benefit from real-time wireless communication and remote access to data and information to support collaborative decision making and problem solving, thereby reducing the need for time-consuming and costly face-to-face meetings. Tedious paperwork can also be reduced by faster and more efficient electronic communication and information flow between the site and the remote design office. Being able to access the right information at the right time can make workers more productive on site. Wireless and mobile technologies can support construction workers on site in the following areas:

- exchanging electronic data with other sites or offices through wireless communications;
- recording site data on a mobile device such as a PDA, Pocket PC or tablet PC;
- instant access to corporate information;
- solving construction problems using project information not available on site;
- virtual meetings with office-based personnel and collaborative decisions on site;
- real-time, on-site integrated audio, video and other data communication.

Recent research suggests that the use of mobile computing and wireless networking on site can reduce time and cost by improving collaboration through information access and recording, and faster communication and information sharing between site personnel and office-based design teams. Areas being supported include videoconferencing, progress monitoring and recording, punch-listing, materials tracking and delivery, access to project drawings and other graphical data, the recording of snagging information, issuing requests for information, and access to technical data for specialised operations such as piling (Rebolj *et al.*, 2000; Saidi *et al.*, 2002; Elvin, 2003; Olofsson and Emborg, 2004; Ward *et al.*, 2004; Bowden *et al.*, 2006).

Planning and implementation of mobile collaboration

The planning and implementation of mobile collaboration in construction highlighted the need for the use of various mobile technologies and wireless communications, collaboration software, and visualisation applications to create a better collaboration environment. It resulted in a new collaboration model integrating these technologies and could potentially improve collaboration platforms and tools used to support design-related tasks on construction sites. Figure 5.6 illustrates the collaboration gap between the design and construction site teams in the traditional paper-based methods, and proposes a mobile collaboration model for the existing inefficient working environment.

The objectives for the planning and implementation of mobile collaboration are:

- to provide a real-time wireless communication platform capable of using collaboration software, visualisation technologies and communication tools to support collaboration between engineers on site and other project members;
- to provide a real-time and mobile telecommunication-based platform capable of sharing visual information to support communication and knowledge sharing between engineers on site and other project members;
- to demonstrate how mobile visualisation can enhance the existing design problem resolution and decision-making processes during the construction stage of a project;

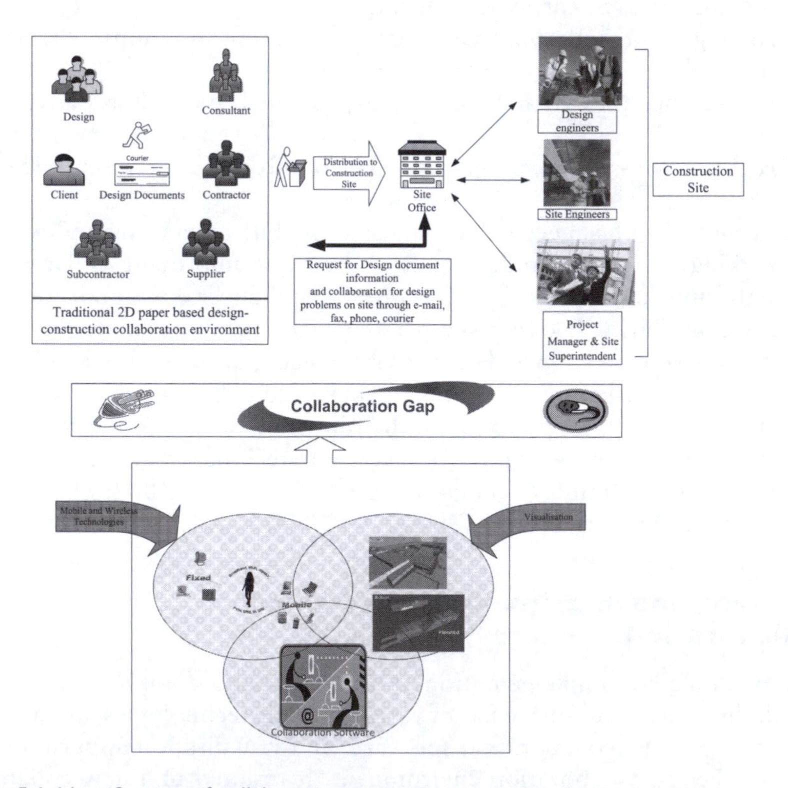

Figure 5.6 Identification of collaboration gap and proposed mobile collaboration.

- to evaluate the use of these technologies within various organisations and identify technical and non-technical issues which occur during the implementation on real construction projects.

The implementation of mobile collaboration is based on the Planning and Implementation of Effective Collaboration in Construction (PIECC) project's framework presented in Chapter 3. A specific framework for the implementation of mobile collaboration technologies has been derived; this is presented in Figure 5.7 and is subdivided into project-, people- and technology-based activities.

Mobile collaboration brief

- Developing a shared vision is an important stage in the planning for a successful or effective collaboration. It should include the scope and objectives, clearly defined requirements of all users/stakeholders, agreed success criteria, and a clear perspective on the role of information and communication technology (ICT). For the success of mobile collaboration, adequate detailed guidance should be provided so that all members of collaboration can stay focused and productive.

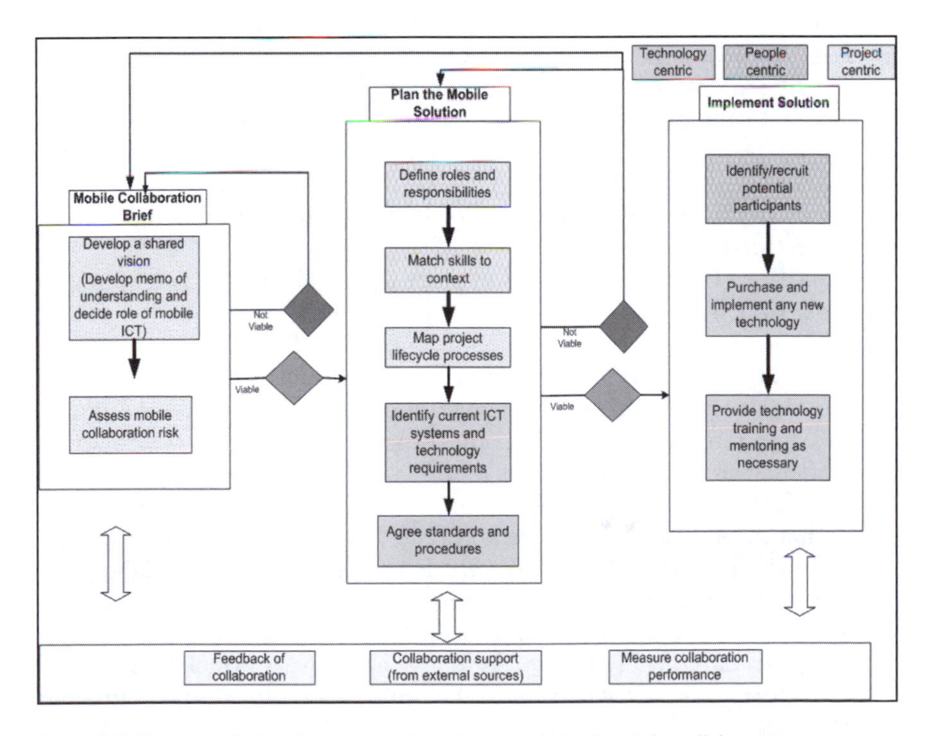

Figure 5.7 Framework for planning and implementation of mobile collaboration.

- Assessing mobile collaboration risk should identify, analyse and communicate risks associated with any activity, function or process in the mobile collaboration implementation. Different assessment approaches can be used depending on the purpose and scope of the collaboration. In this case, the assessment approach should identify risks before the mobile collaboration begins and try to predict what their effects will be.

Plan the mobile solution

- Roles and responsibilities should be well defined as people perform better when they know what is expected of them. It is also important for all those involved to know what they can and should expect from each other.
- Having identified the roles and responsibilities required for mobile collaboration the next step is to break these roles into individual tasks. It is then essential to identify the right people with the relevant skills to complete the tasks. Skills of people and the technologies required should be decided together to implement a mobile collaboration solution. It is pointless having highly skilled individuals using basic tools, or highly sophisticated tools for use by unskilled users.
- If a construction organisation is not explicit about the processes and systems to be used in collaboration, it is certain that individuals will use different systems to communicate (email, fax, face-to-face meetings etc.) with the potential for some of them to be inefficient. It is best to model the business in enough detail to capture the key activities and processes of an organisation. Then a new model that integrates processes and mobile technologies in order to create a better collaboration environment is developed.
- Identify ICT systems currently in place to establish what is required in order to deliver the mobile collaboration.
- The basis for collaboration is communication; therefore it is necessary to define rules and protocols for communication. These should be in the form of information exchange and communication procedures, and standards for structuring and representing project information. Information standards and procedures should cover the production of information within ICT systems, the formatting and structuring of that information, exchange and reuse. One way of identifying such standards and procedures is to use the Avanti approach to collaborative working in construction. The Avanti information standard and procedure is at the heart of the approach. It provides a common standard and procedure for use by project teams, and covers principles for sharing information, drawings and schedules in an agreed and consistent manner.

Implement solution

- Using the information generated in the 'define roles and responsibilities' and 'match skills to context' processes from the PIECC model, the potential participants in the mobile collaboration solution can be identified. This should be done by examining the skills of existing personnel in the organisations. If the right skills are not available in the organisation then it is worth considering recruiting new personnel.
- Having established the technology needs, it is important to identify the potential costs and long-term benefits associated with the implementation of ICT. Decisions can be made about how the cost of new technology is divided among the project team members. The implementation of a new ICT system, in the form of either hardware or software, will require the physical implementation of the system and associated user training and mentoring. This should cover the maintenance of the system and usage. Generally, in training there is a lack of concern about how software systems can be better implemented to collaborative working processes within the organisation.
- Training requirements for all participants in mobile collaboration should be planned according to the skill and abilities of users, and technologies to be used. The cost and time involved in training participants to enable them to use the technology for collaboration should not be underestimated. The cost of training for ICT use in collaborative environments can be a major part of the overall cost of mobile collaboration. Therefore, it may be helpful to obtain feedback from users and ICT managers from previous collaborations to better capture the user's training requirements for mobile collaboration.

Evaluation of mobile collaboration

- Employees will encounter regular changes in their way of working when involved in the new mobile collaboration environment; therefore the mechanisms for obtaining feedback through observations, interviewing and workshops must continue throughout the life of the collaboration.
- There may be a need to bring external expertise for this process from outside the collaboration venture and the organisations involved to ensure objectivity and cover the variety of skills.

Implementation scenarios

A scenario provides a context for the representation of a set of complex factors and variables that affect a given situation in which a series of 'what if' options need to be examined (Schwartz, 1999; Karlson *et al.*, 2003). Scenarios can represent a long-term view of uncertainties to explore the way the world might be in the future. They are not predictions; they just present possible states of the future, unlike business forecasting or market research (Schwartz, 1999).

Traditionally, scenario planning was used to improve the quality of decision making. However, in recent years interest has moved from developing scenarios to successfully using them, which depends on the ability to look at future conditions (Ringland, 2002). Scenario planning was used in this research after reviewing the literature and identifying the collaboration requirements for construction through a questionnaire survey (Koseoglu and Bouchlaghem, 2008). Following the development of the scenarios, validation meetings were held with experts from a number of organisations in the construction industry to ensure that these scenarios were applicable and appropriate within the construction environment.

Three different scenarios are therefore presented in this chapter to represent the use of technologies for communication and collaboration between the construction jobsites and design teams. These scenarios are titled Mobile 2D/3D, 4D Collaboration and 3G Communication. They include combinations of various mobile devices, wireless networks and visualisation technologies. Figure 5.8 briefly presents the key technologies underlying the scenarios and the architecture of wireless networks which constitutes

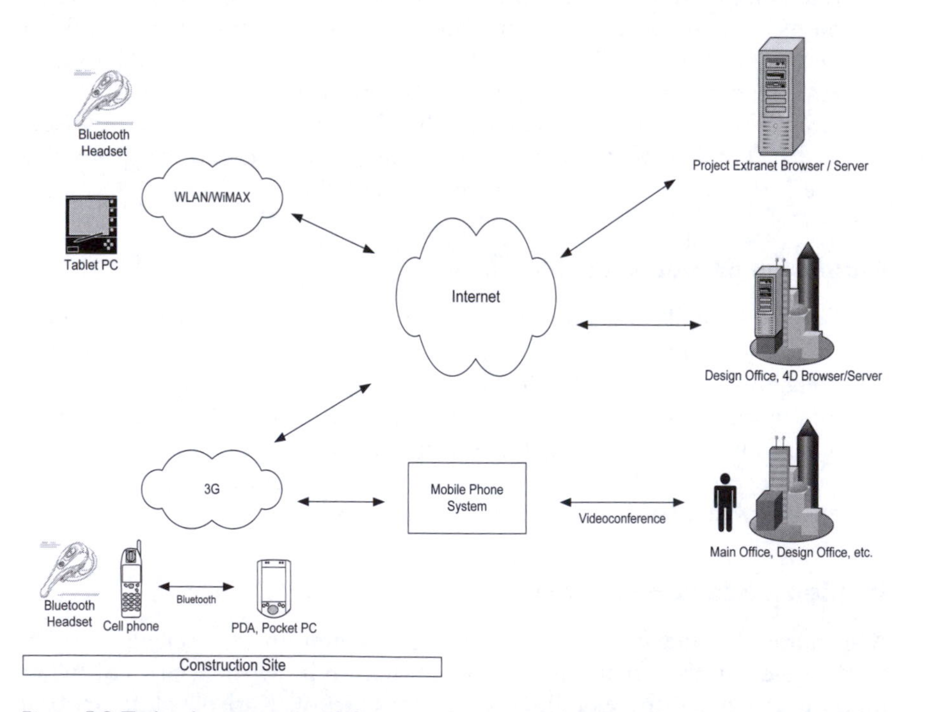

Figure 5.8 Technology set-up: mobile communication architecture for data, audio and visual information exchange.

the main communication set up on site. The scenarios development and the architecture of wireless networks were based on existing literature and technology reviews, and the needs of the construction industry identified through the questionnaire surveys (Koseoglu and Bouchlaghem, 2008).

The objectives of the scenarios for the planning and implementation of mobile collaboration are:

- to provide a real-time wireless communication platform capable of using collaboration software, visualisation technologies and communication tools to support collaboration between engineers on site and other project team members;
- to provide a real-time and mobile telecommunication-based platform capable of sharing visual information to support communication and knowledge sharing between engineers on site and other project members.

Figures 5.9, 5.10 and 5.11 show the infrastructure of the wireless networking systems used in the scenarios for communication and collaboration between the construction and design teams.

Table 5.2 explains the proposed technologies for each scenario together with their main benefits and possible drawbacks; this was used as a guide for the technology set-up within the scenarios explained in the following sections.

Mobile 2D/3D scenario

This scenario illustrates the case of a site engineer who communicates and collaborates with the project team in real time, accessing project drawings, documents and specifications, and requests information for design queries or buildability problems. Figure 5.12 shows a scenario based on the

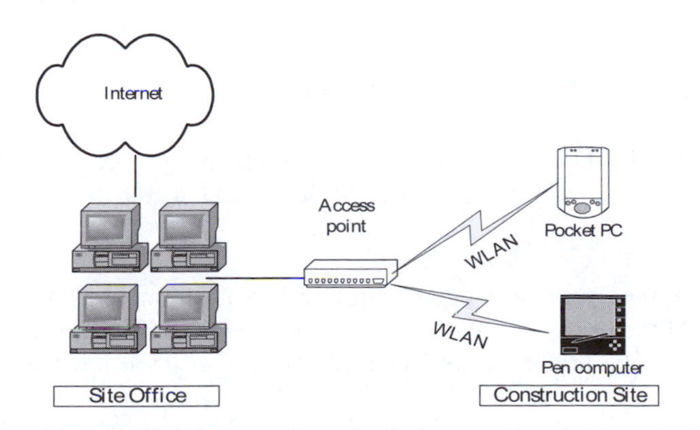

Figure 5.9 WLAN infrastructure at construction site.

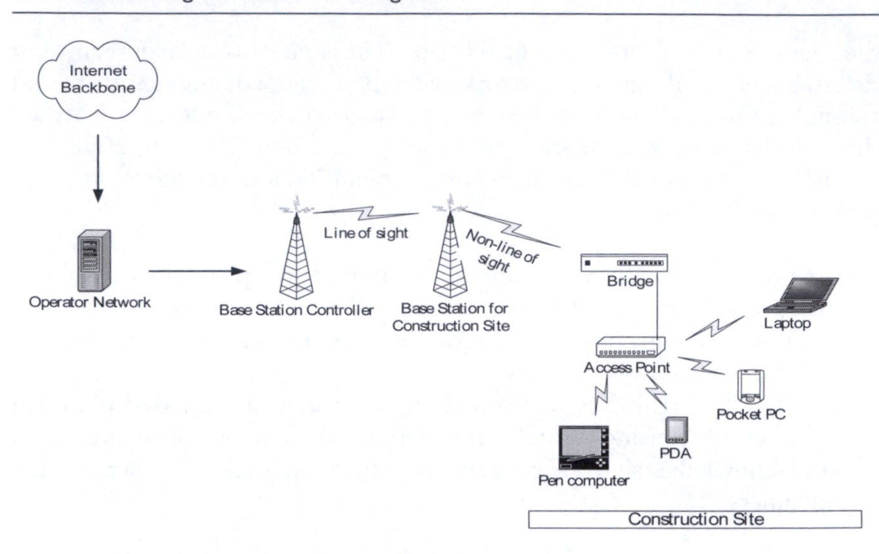

Figure 5.10 WiMAX infrastructure at construction site.

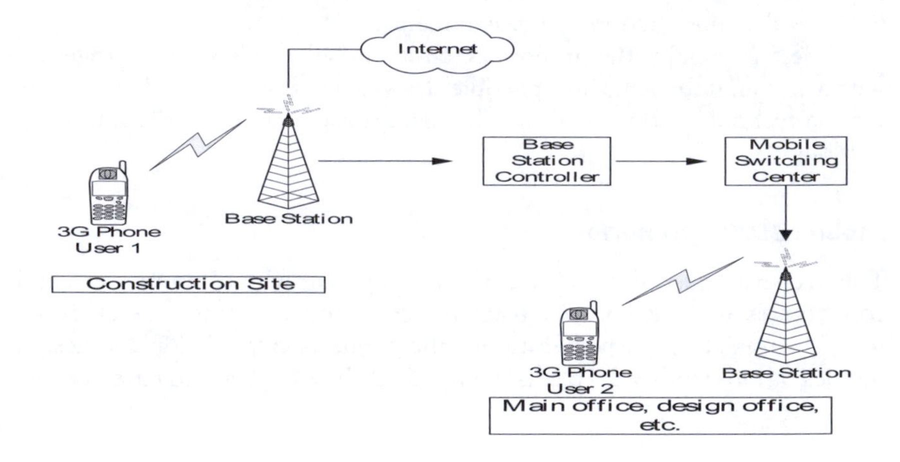

Figure 5.11 3G mobile phone system architecture.

exchange of 2D/3D design documents between the project team members and the construction jobsite. The mobile 2D/3D scenario is described as follows:

- The engineer uses a rugged (suitable for outdoor and harsh environments) Bluetooth- and WLAN-enabled tablet PC, and Bluetooth-enabled headset. The tablet PC and headset are linked through the Bluetooth network.
- The construction site is covered with a wireless network (802.11g) using outdoor access points and antennas. It is likely that, in the future,

Table 5.2 Technologies proposed for use in the scenarios

Technology	*Scenario and process identified*	*Benefits*	*Drawbacks*
Tablet PC (device)	Mobile 2D/3D (accessing published drawings, documents and specifications) 4D collaboration (monitoring progress on site)	Similar to a small laptop, lightweight design holds like a book and can be used sitting or standing Suitable for viewing large-format information such as published 2D/3D drawings Long battery life provides for using in different environments for long time No barrier between the user and others, information can easily be shared with a person nearby User can use own handwriting on the screen. This leads to drawing figures, sketching and using the PC as a worksheet Gives the user freedom of movement Most have integrated Bluetooth and WLAN technology in order to be used in wireless networks	Requires two hands to operate on account of larger form than small devices Rugged ones have to be used at construction sites owing to harsh working environments; these are more expensive than normal tablet PCs
Pocket PC (device)	3G communication (instantaneous decision making, sharing visual information and knowledge)	Pocket PCs have large and clear full-colour screens compared with personal digital assistants (PDAs). Manufacturers produce devices which have Bluetooth technology, WLAN and 3G technology Enables capturing videos and photos related to construction on site Real-time videoconference with project team members using 3G networks	Unsuitable for viewing large-format information such as drawings

Table 5.2 continued

Technology	*Scenario and process identified*	*Benefits*	*Drawbacks*
3G mobile phone (device)	3G communication (instantaneous decision making, sharing visual information and knowledge)	Enables capturing videos and photos related to construction on site Real-time videoconference with project team	Limited screen size
Bluetooth headset	Mobile 2D/3D (accessing published drawings, documents and specifications) 4D collaboration (monitoring progress on site) 3G communication (instantaneous decision making, sharing visual information and knowledge)	Hands-free voice calls through VoIP (Voice over Internet Protocol) by accessing Bluetooth-enabled mobile devices Easy set-up and efficient use for mobile workers on site	Needs to be protected from harsh outdoor environments
WLAN (wireless local area network)	Mobile 2D/3D (accessing published drawings, documents and specifications) 4D collaboration (monitoring progress on site)	No additional data transfer cost apart from installation and hardware Enables transferring large files thanks to higher data transfer rates than 3G Information can be downloaded whenever required, limiting the need for device memory	Requires installation of WLAN access points and antennas (limited coverage) Construction site requires outdoor hardware components which are more expensive than indoor WLAN equipment

WiMAX (Worldwide Interoperability for Microwave Access)	Mobile 2D/3D (accessing published drawings, documents and specifications) 4D collaboration (monitoring progress on site)	Deliver broadband connectivity in a larger geographic area than WLAN and enable greater mobility for high-speed data applications Cell radius of 4–6 miles and range up to 30 miles	Set up like mobile phone system by using base stations that serve a radius of several miles and more expensive than WLANs Certification has been delayed and unlicensed network equipment are used by network providers WiMAX-compatible mobile devices have not been produced yet Regular service provider fee besides hardware and installation costs
3G (third-generation mobile phone system)	3G communication (instantaneous decision making, sharing visual information and knowledge)	Video on demand, video conferencing, high-speed multimedia and mobile internet access are the important features Data transfer rates are much faster than the existing mobile telecommunication networks Provides wider area coverage than WLAN and WiMAX	3G licences have been awarded around the world at huge costs by many governments and operators still offer expensive prices for high rates of use Data transfer rate is not suitable for downloading large files such as models and drawings Network coverage needs to be expanded at remote sites
Bluetooth	Mobile 2D/3D (accessing published drawings, documents and specifications) 4D collaboration (monitoring progress on site) 3G communication (instantaneous decision making, sharing visual information and knowledge)	Provides fast and secure voice and data communications Low-cost wireless network solution Cable replacement Suitable for personal area networking	Short range and limited data transmission rate Range up to 10 metres and theoretical data transfer rate 1 Mbps

Table 5.2 continued

Technology	*Scenario and process identified*	*Benefits*	*Drawbacks*
Project extranet (collaboration software)	Mobile 2D/3D (accessing published drawings, documents and specifications) 3G communication (instantaneous decision making, sharing visual information and knowledge)	Real-time access to published drawings and model files as 2D or 3D, project documents, specifications etc. Better drawing and document circulation between project members Provides storage space for visual information such as videos and photos captured from the construction site for sharing within project team Notify users through emails if any change or mark-up raised during the project	Slower rates of data transmission because of poor web server
Web-based 4D modelling software	4D collaboration (monitoring progress on site)	Real-time access to updated schedule and 3D model of the construction Monitoring the resources, cost etc. during the actual construction according to planned programme	Technical problems during web-based collaboration due to internet infrastructure

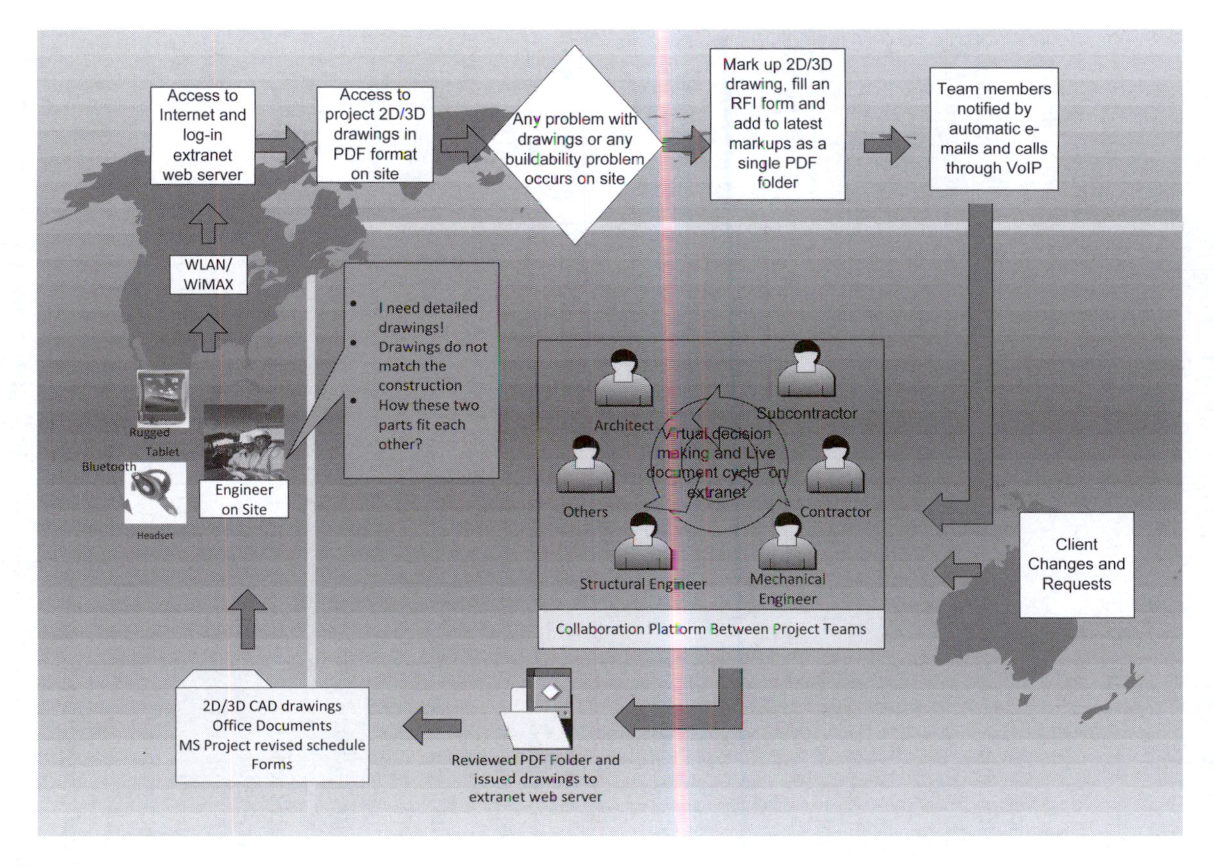

Figure 5.12 Mobile 2D/3D scenario.

operators will be able to provide WiMAX internet access to subscribers without the need to install any antennas and access points.

- The engineer connects to the web using the wireless network and logs in to the web server of the project extranet service provider.
- Audio communication between the project members is performed through Skype, a free VoIP service provider over the web.
- After connecting to the project extranet and Skype account, project members can access the server and then search for and display any required 2D/3D drawings or specification documents in portable document format (PDF).
- Publishing the drawings in Adobe Acrobat Professional and Acrobat 3D formats enables the project team to read and comment on any type of drawing or MS Office document without the need to access CAD or project management software. 2D and 3D CAD files produced by different parties in different formats can be easily converted to a standard format for sharing the CAD data, such as PDF.
- The site engineer can monitor the construction process by displaying drawings and other relevant documents on a tablet PC screen. In case of queries arising, the PDF files can be marked up by hand using the pen-based tablet PC and then a request for information (RFI) can be raised and attached as a 'mark-ups' folder. The RFI can be put together in the form of an intelligent PDF folder which contains multiple files such as CAD, Office and MS Project documents and images.
- Project team members are then notified by email each time a document is uploaded to the collaborative environment. In addition, the site engineer has the option to notify relevant parties through the VoIP.
- A web-based collaborative decision-making and document exchange platform is available for the designers, client, contractor, sub-contractors and so on. If a problem occurs on the construction site or the client requests a change to the design, then a redesign can take place in real time and be communicated back to the site engineer.
- Revised drawings, updated MS Project schedules and even cost changes included in an Excel file can be inserted into a single PDF folder and uploaded to the project server.
- All project team members are notified about changes or revisions to the project documents. The revised files can also be sent to the construction site as they are produced.

4D collaboration scenario

This scenario illustrates the case of a site engineer who checks and monitors the progress of the construction process against planned schedules using a collaborative 4D modelling platform and communicates with the rest of the project team using a 3D model with built-in resources, schedules, tasks and costs to explore various 'what if' scenarios on the construction jobsite. Figure 5.13 presents the scenario, which is described as follows:

- Similarly to the previous scenario the engineer uses a rugged Bluetooth- and WLAN-enabled Tablet PC, and Bluetooth-enabled headset.
- Wireless communication is established using outdoor access points and antennas or (in future scenarios) operators may provide WiMAX internet access to subscribers without the need for installing any antennas and access points.
- The engineer connects to the web through the wireless network and accesses the 4D (3D + time) model on the server where the audio communication between project members is established through Skype.
- The engineer displays the 3D model of the building, Gantt chart and resources (material, labour and equipment) on the screen of the Tablet PC at the same time as other project team members. They can all view the construction activities, schedule and resources using the 3D model, making it easier to follow the different stages of construction.
- Using a real-time and web-based 4D modelling platform in a collaborative environment enables the project members to have an interactive visual representation of resources, changes in schedule and design. The site engineer and project team can then visualise different 'what if' scenarios when buildability issues arise, making the decision-making process more collaborative, faster and easier.
- Real-time updates on the project schedule and resources are implemented following the changes in design. This enables a better coordination with sub-contractors and suppliers. Feedback from the site engineer in the case of any actions which result in a change to the plans, drawings, specifications, materials and so on makes it easier to update other related information and leads to better project coordination.

3G communication scenario

This scenario illustrates the case of a site engineer who makes decisions and shares information and knowledge to solve buildability problems on a construction site using 3G visual communications. Figure 5.14 presents the scenario and its details are explained below:

- The engineer has a Bluetooth-enabled 3G smart phone and a Bluetooth headset which provides hands-free voice communication that isolates external sound during a videoconference on a construction site.
- While the site engineer monitors construction, buildability problems or unexpected difficulties due to weather, soil conditions or failures of materials may occur. 3G provides the opportunity of sharing visual information, accessing tacit knowledge from experts and making

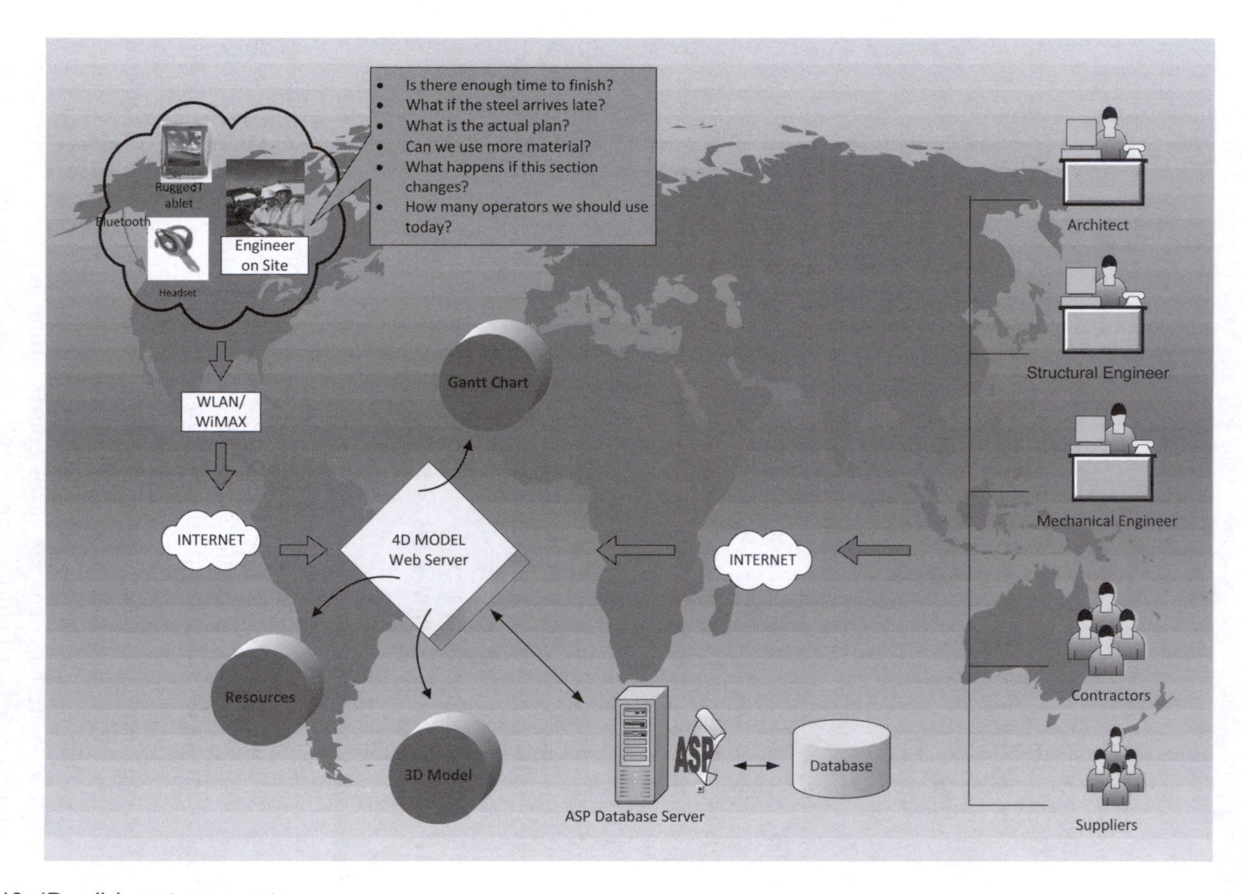

Figure 5.13 4D collaboration scenario.

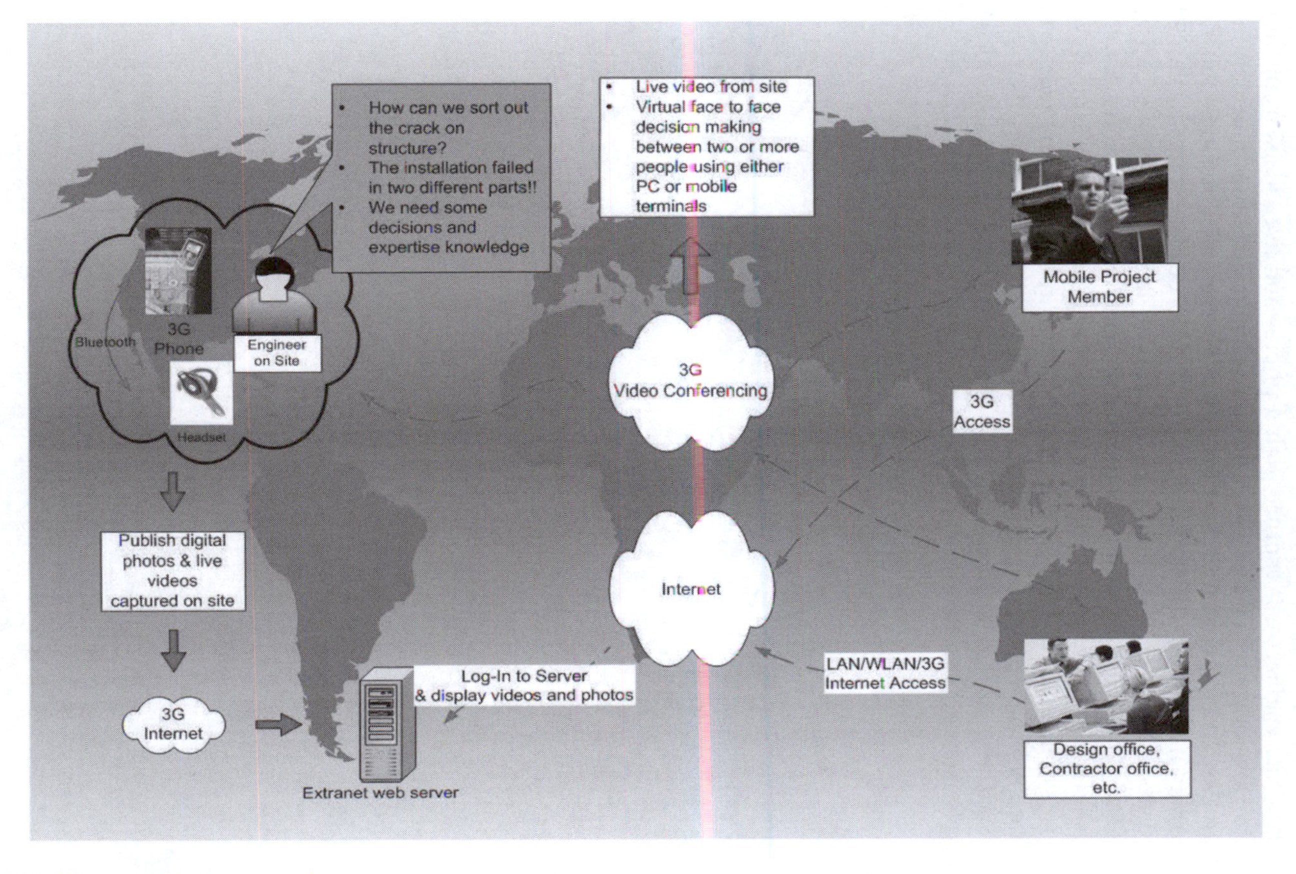

Figure 5.14 3G communication scenario.

collaborative decisions using an interactive communication platform with other project members in real time.

- The site engineer can capture digital photos and live videos from the construction site and using internet access can upload this information to the extranet web server or directly send it to a project team member.
- Apart from image distribution and live video delivery, the 3G phone system allows users to make virtual face-to-face video phone calls. Owing to the outdoor and harsh environments of construction sites, it is difficult to have real-time videoconference meetings in the same way as is done indoors. However, 3G provides any time–anywhere face-to-face meetings, and site engineers can participate in a video conversation with both 3G mobile phone and PC users who have 3G cards and video telephony software.

Chapter 6

Multidisciplinary Collaborative Design Optimisation

Fan Yang and Dino Bouchlaghem

Introduction

In building design, conceptual options are refined through an iterative process that seeks to produce a proposal in which the combined variables satisfy a set of collective criteria managed by various design professionals concurrently. Hence there is a need for design tools that can help designers better manage this collaborative design development process. It is well accepted that performance improvement in building design cannot just be based on single disciplines in isolation. Therefore, better, system-orientated, holistic, multidisciplinary approaches to building design are needed to improve design quality (NSF, 1996). This chapter discusses multidisciplinary collaborative design with a focus on the process used to determine the option (from a set of alternatives) that best meets all design criteria. This Multidisciplinary Collaborative Design Optimisation (MCDO) process divides a single system into a group of smaller sub-systems to effectively manage interactions between them, and the search for and evaluation of alternatives. In the context of building design, the system refers to building design as a whole, with sub-systems representing the various design disciplines or parts of the building such as the different spatial zones. Such an approach could reduce the time and cost associated with the multidisciplinary design cycle. The chapter also presents a Pareto Genetic Algorithm-based Collaborative Optimisation (PGACO) framework that supports interactions between multiple disciplinary tasks and coordinates conflicting design objectives. The use of a Pareto-based genetic algorithm (GA) approach helps to mitigate the problems that usually occur in multi-objective searches that can result in multiple solutions rather than a single optimum value.

Collaborative design

Fragmentation and specialisation in the construction industry means that no single professional has all the knowledge needed to take full control of the design of a complex facility. Collaboration between different disciplines is essential for the success of building design. In order to support collaboration, design models must provide participants with functions for information sharing, task coordination and conflict resolution. With the increasing capabilities offered by computers as communication devices and the use of dedicated collaborative environments, the sharing of information and interaction between designers has been greatly improved. Collaboration systems provide access to design information in various formats and support communication between multidisciplinary design team members.

Furthermore, the effective management of conflicting requirements and trade-offs between variables during the design process is essential if collaborative design is to deliver a successful product. One type of conflicts stems from differences in semantics and data standards, whereas others result from the interdependencies between discipline-specific design criteria. Traditional methods to manage these conflicts are based on negotiations and face-to-face meetings, which can be time consuming and limited to the knowledge and experience of those directly involved.

In many ad hoc design procedures, individual design teams are assigned sub-sets of the design variables, individual parts of the analysis, and local design objectives that are only vaguely aligned with the overall goals of the design. Sequential uncoordinated decisions by different design teams lead to a non-integrated process and may reach a design solution that is not optimal for the whole system. Decisions within the design sub-systems are handled by discipline experts (such as structural or mechanical and electrical designers), who have specialised knowledge and use domain-specific analytical software.

Optimisation methods that enable designers to consider a large number of options and assess the extent to which they satisfy the different criteria have proved a successful alternative in handling conflicting requirements (Bouchlaghem, 2000). However, some of the characteristics specific to building design make the application of optimisation techniques challenging; such characteristics include the large number and different types of variables, which can result in high computational demands, and the varying effects that design variations tend to have on different design criteria. MCDO is regarded as an effective method for the following reasons (Braun and Kroo, 1995):

- *Decrease in computational burden*: the solution time and computational cost for most analysis and optimisation increase at a high rate with the increase in size and complexity of the design problem. MCDO enables the exclusion of the local discipline-specific variables at system level. This means that the individual sub-system (discipline-focused) designs are, in a sense, hidden at system-level decision making. Therefore the workload and communication requirements of the system-level coordination process are significantly reduced, whereas the sub-system optimiser is the responsibility of discipline-specific designers without consideration of influences from other disciplinary designs.

- *Good fit within conventional organisations*: large design projects involve a number of participants from various design disciplines, usually in separate locations. Disciplinary decomposition should be well suited for use in conventional design settings. The MCDO principles are in line with the traditional design process whereby a design team leader (system-level decision maker) is responsible for the overall design objectives while managing a set of discipline experts (sub-system optimiser) into agreement.
- *Does not require analysis software integration*: the salient feature of collaborative design is the large and monolithic analysis and design codes, making the integration of engineering assessments become complex as the number of disciplines increases. MCDO allows designers to use discipline-based applications independently.
- *Autonomy of discipline-based designs*: MCDO allows more design freedom when making disciplinary decisions while enforcing multidisciplinary compatibility at system level. This means that the specialised designer is free to specify not only the discipline-specific design variables but also those that are outputs from the analysis of another design area.

Systems approach to collaborative building design

Given the large number of design variables and complex relationships between them, it is difficult for one person or team to handle the entire set. Close interaction between disciplines is required as decisions made within one affect how the whole system responds to the changes required in another. Hence appropriate methods are chosen to manage these interactions and to achieve a suitable overall solution. A system-based approach helps to achieve this through the partitioning of the complex system into smaller and more manageable components that are logically linked to defined objectives using systematic procedures (Al-Homoud, 2005).

Applying a system approach implies the implementation of a decomposition and coordination strategy, the underlying principle of collaborative design optimisation. The system approach is necessary to solve complex design problems for the following reasons (Papalambros, 2002; Al-Homoud, 2005; Choudhary and Michalek, 2005):

- It is a systematic, logical design procedure whereby conflicts between sub-systems are effectively solved through the adaption of gradient-based optimal sensitivity (Braun and Kroo, 1995) or artificial intelligence technology (Anumba *et al.*, 2003; Chen *et al.*, 2005).
- Design variables increase significantly if a new sub-system is added in the design. A decomposition scheme allows each sub-system to consider

only the sub-system-specific variables. This improves the overall system's capability in handling variables, compared with the traditional sequential design approach.
- Interactions between different subsystems are identified and considered before carrying out the design, which reduces iterations in the design process and the overall design cycle.

These decomposition and coordination methods have been used in building and infrastructure design in different ways:

- Balling and Rawlings (2000) used a two-level optimisation framework in bridge design. In this case, bridge design was decomposed into superstructure and deck groups as sub-systems, whereas the system level is responsible for the overall design objective of cost minimisation and coordination. The coupling variables (outputs from one discipline used as inputs in another) originate from the interactive force between the cable and the deck of a typical suspension bridge. The system-level optimiser sends the target values of coupling variables to the corresponding sub-systems; when the target values match the actual values that satisfy the constraints at sub-system level, then a consistent overall design is found.
- Choudhary and colleagues (2005) developed a multilevel optimisation framework for building performance analysis using an analytical target cascading (ATC) method. This framework was based on component decomposition whereby a complex healthcare facility was divided into six function zones. Every zone was then further partitioned into several rooms and cubicles hierarchically. Every part partitioned was regarded as a decision model linked to the relevant analysis software. Linking variables represented decisions shared between two or more decision models at the same level. The vertical relationships between decomposed levels were embodied through building performance targets. The ATC framework coordinates multiple decision-making tasks to reach a compatible solution, and provides explicit decision support including trade-offs between performance targets and the use of simulation tools in decision-making.
- Pushkar and colleagues (2005) introduced a methodology for the optimisation of environmental design based on the sequential decomposition that was usually used in the chemical and manufacturing industries. Through the proposed methodology, the entire design variables were divided into a construction group, an operational energy group and a maintenance to demolition group according to the extent of their relative environmental impact. In such a manner, environmental optimisation could be performed within each group separately, and then the partial decisions combined. Finally the overall environmentally optimal solution for the entire building was obtained.

The one limitation of the above methods is that they seek to improve building design in one performance area (e.g. energy performance or structural design). An optimal design solution that satisfies all design criteria greatly relies on the cooperative efforts of a group of professionals, including architects, structural engineers, mechanical and electrical services engineers, construction engineers and quantity surveyors. The other limitation is that building design in the above applications is not dealt with as a multiobjective optimisation problem. Multiobjective optimisation poses big challenges whereby different design disciplines must be integrated so that they are consistent with one another while deciding on properties that, in combination, achieve the best performance of the overall system (Choudhary and Michalek, 2005).

Multidisciplinary Collaborative Design Optimisation

Many of the fundamental characteristics of MCDO are reflected in its multilevel approach. Figure 6.1 illustrates the links between the system optimiser and the sub-systems. The variables of the original design problem are expressed in terms of local X_i, shared X_s and coupling Y_{ij} variables (see Table 6.1 for explanation of these) according to the discipline partitioning. The auxiliary variable X_0 is used as a new variable to replace the shared

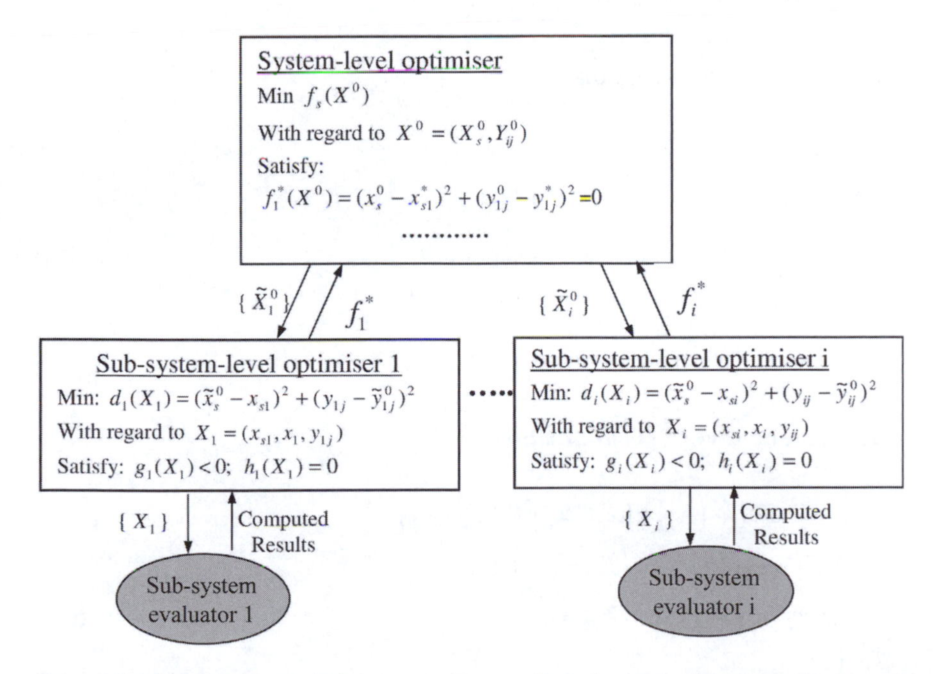

Figure 6.1 Multidisciplinary Collaborative Design Optimisation architecture (Braun and Kroo, 1995).

Table 6.1 The main type of variables in the MCDO problem

Type of variable	*Definition*	*Examples in building design involving structural and HVAC disciplines*
Shared variable	The type of variables that influence more than one disciplinary design	Size of window
Coupling variable	Output from one discipline used as input in another	Dimension of walls, floors and roof (from structural to HVAC discipline)
		Weight and position of heating and cooling equipment (from HVAC to structural discipline)
Local variable	The type of variables that are handled by an individual discipline/designer; their values do not affect other disciplines	Dimension of column
		Supply air rate

HVAC: heating, ventilation and air conditioning

X_s and coupling Y_{ij} variables at system level. Such auxiliary variables make the links between disciplines to enable them to develop the design concurrently. In addition, compatible constraint functions ($d_i^*(X^0)$) are introduced to ensure consistency in the design process. The detailed formulation of system and sub-system formulation is presented in equations 6.1 and 6.2 and illustrated in Figure 6.1.

System-level formulation

$$\text{Minimise } f_s\left(X^0\right)$$
$$\text{With regard to } X^0 = \left(X_s^0, Y_{ij}^0\right)$$
$$\text{Satisfy: } d_1^*\left(X^0\right) = \left(x_s^0 - x_{s1}^*\right)^2 + \left(y_{1j}^0 - y_{1j}^*\right)^2 = 0 \qquad (6.1)$$
$$\ldots$$
$$d_1^*\left(X^0\right) = \left(x_s^0 - x_{si}^*\right)^2 + \left(y_{ij}^0 - y_{ij}^*\right)^2 = 0$$

Sub-system-level i formulation

$$\text{Minimise: } d_i^*\left(X_i\right) = \left(\tilde{x}_{si}^0 - x_{si}\right)^2 + \left(y_{ij} - \tilde{y}_{ij}^0\right)^2$$
$$\text{With regard to } X_i = \left(x_{si}, x_i, y_{ji}\right)$$
$$\text{Satisfy: } g_i\left(X_i\right) < 0; h\left(X_i\right) = 0 \qquad (6.2)$$

Where

$f_s\left(X^0\right)$ is the objective function in the system level;

X^0 is the vector of system-level variables;

x_s^0 is the vector of the shared variables in the system level;

y_{ij}^0 is the vector of the coupling variables in the system level;

x_{si}^* is the vector of optimal value of shared variables sent by the *i*th sub-system;

y_{ij}^* is the vector of optimal value coupling variables sent by the *i*th sub-system;

$d_i^*\left(X^0\right)$ is the *i*th compatible constraint in system level;

$\tilde{X}_i^0$ is the vector of target value of system-level variables that are sent to the *i*th sub-system;

$d_i\left(X_i\right)$ is the objective function in the *i*th sub-system;

X_i is the vector of decision variables in the *i*th sub-system;

x_i is the vector of local variables of the *i*th sub-system;

x_{si} is the vector of shared variables corresponding to the *i*th sub-system;

y_{ji} is the vector of coupling variables that is input to the *i*th sub-system but is output from the *j*th sub-system;

$\tilde{x}_{si}^0$ is the vector of target value of shared variables sent from system level to the *i*th sub-system;

$\tilde{y}_{ij}^0$ is the vector of target value coupling variables sent from system level to the *i*th sub-system.

In the system level, the optimisation handles the interactions between disciplines and adjusts the target values of auxiliary variables (X_s^0, Y_{ij}^0), seeking to optimise the system objective function $f_s(X^0)$. The optimiser for sub-system *i* receives the system target values pertinent to sub-system *i*, and minimises any discrepancy between the system target values (i.e. $\tilde{X}_i^0$) and their corresponding sub-system values (i.e. x_{si} and y_{ij}) within the *i*th sub-system, subject to satisfying the sub-system's local design constraints (i.e. g_i and h_i). The compatible constraint functions at system level are the same as the objective functions at sub-system level (i.e. $d_i^*(X^0)$). Values of interdisciplinary variables are agreed between disciplines when the compatibility constraints reach zero. Although the sub-systems' objective functions take the same form as the compatibility constraint functions of the system level, both x_{si} and y_{ji} are variables in the ith sub-system level, while both x_s^0 and y_{ji}^0 are variables at the system level. In other words, both objective functions of sub-system level and constraints functions of system level are similar but with different variables that have the same physical meaning.

In the MCDO formulation, each sub-system optimisation is given sufficient degrees of freedom to achieve a design that is feasible with respect to its local constraints, because the sub-system-level optimisers are able to manage all discipline-specific variables (X_i) which include the local x_i, shared x_{si} and coupling y_{ji} variables. It should be noted that the target values of auxiliary variables corresponding to the ith sub-system (i.e. $\tilde{x}_{si}^0$) remain constant during the sub-system optimisation, while the optimal values of auxiliary variables sent from the sub-system optimisers (i.e. x_{si}^*) are fixed parameters during the system optimisation.

The PGACO framework development

Challenges in building simulation integration and systematic design motivated the need to explore the application of MCDO in building design. As explained above, MCDO is a methodology that supports design as a process and provides a collection of tools and methods that permit the trade-off between different disciplines. Furthermore, it could be regarded as a concurrent engineering design tool for large-scale complex systems that can be affected by the optimal design of several smaller functional units or sub-systems (McAllister *et al.*, 2005). This section presents a typical building design problem in the context of MCDO and describes the process of developing the PGACO framework.

The two-level multiobjective collaborative optimisation (MOCO) framework developed by Tappeta and Renaud (1997) provides solutions that integrate interrelated simulation software and coordinate conflicts between the various disciplines. In the MOCO approach, all variables are firstly grouped into three types: shared, coupling and local variables. Table 6.1 defines these types of variables and identifies how they apply to a building design example. The reader is encouraged to refer back to this table to put an instance of a physical meaning on the variables.

In addition, auxiliary design variables are introduced to replace the shared and coupling variables in order to process more than one discipline-based optimisation. The values of these auxiliary variables are constant and are called target values during the sub-system-level optimisation, although they could be adjusted by the system-level optimiser. The sub-system-level optimiser can also adjust the value of these shared and coupling variables; thus a situation will emerge in which the same variable at the system level receives different values from each sub-system. In order to achieve consistency in the design process, compatibility constraints are introduced at system level. These constraints are the sum of the square differences between the values of the auxiliary variables and their optimal values in the corresponding sub-system level. If these constraints are set to zero, it will imply that a compromised multidisciplinary design solution is obtained.

Reasons for adopting a Pareto-based genetic algorithm

The limitations of the MOCO framework have been analysed by Yang (2008) considering the following aspects: step size, local optimum, approach of multiobjective problem and delinquent nature of the MOCO formulation. In order to overcome these limitations, a Pareto-based GA at the system level is used. The reasons for this are:

- *Global optimisation algorithm*: the genetic algorithm is best known as a global search optimisation. If this algorithm is used for the system-level optimisation, both system- and sub-system-level processes avoid being trapped in a local optimum, even if a gradient-based method is used at the sub-system level. This is because the target values of shared and coupling variables returned by the system level influence the sub-system-level optimisation through the objective function used at sub-system level.
- *Multiobjective approach*: in the MOCO framework, the weighting method is used to solve multiobjective problems. This approach requires the introduction of weighting factors for the different objectives before performing the optimisation. Values of these factors are subjective and mostly depend on the client; hence they are difficult to fix because of the unexpected changes that often occur during the process of building design. For example, a client may set a higher value for the weighting factor of the running cost than that of capital cost and then change these at a later stage of the process. However, the Pareto method can provide several optimal solutions with different sets of weighting factors for the same problem.
- *Delinquent nature of the MOCO formulation*: the MOCO formulation can result in the Karush–Kuhn–Tucker (KKT) conditions not being met; consequently the process could stop at any point without reaching an optimum solution (Alexandrov and Lewis, 2002; Lin, 2004). Most gradient-based methods use KKT as a terminal criterion, such as sequential quadratic programming. This limitation can be avoided by using GA-based methods and setting the termination criterion as a predefined number of generations.

The above explain the advantages of using a Pareto-based GA at the system-level optimisation in the MOCO framework. However, this algorithm can also have some limitations in handling constraint functions. For example, when penalty function strategies are adopted to move from a constrained to an unconstrained problem, difficulties arise when the fitness assignment is based on the Pareto rank of a solution and not on its objective function value (Jimenez *et al.*, 2002). Therefore this study proposes a two-cycle framework, internal and external, whereby the MOCO's type of system-level optimisation can be regarded as an unconstrained problem. A detailed description of the proposed framework is presented in the next section.

The conceptual PGACO framework

The PGACO framework consists of two cycles, internal and external. The internal cycle operates at two levels: system and sub-system. The internal cycle aims to adjust the stochastic values of interdisciplinary variables (shared and coupling) to bring them to within the feasible range through a two-level framework. The process of the Pareto-based GA is performed in the external cycle; this determines the original design objective functions (e.g. minimise capital cost) and then generates new stochastic values of interdisciplinary variables to send back to the internal cycle for the next-generation run.

Here 'stochastic' means that the values are chosen within the constraint bounds at random; 'feasible' means that the value of variables can satisfy all design constraints (e.g. allowance for shear force of beam). The processes used in the two cycles are explained below.

Internal cycle of the PGACO framework

In the internal cycle, auxiliary design variables (i.e. X^0) are introduced to replace the shared and coupling variables. An important feature in the PGACO framework is that a compatibility objective function

$$\sum_{i=1...n} d_i^*(X^0)$$

is introduced at the system level to ensure consistency in the optimisation process. If this objective function is minimised to zero, a compatible solution is obtained. Here 'compatible' means that the same variables are returned with equal values at both system and subsystem levels. The mathematical formulations of system and sub-system levels are presented in Figure 6.2.

As shown in Figure 6.2, the internal cycle starts when receiving the stochastic values of interdisciplinary variables, these are then sent to the sub-system level. After completing the optimisation process within each sub-system, the optimal values of interdisciplinary variables (X_{sh}^{1*}, Y_{1j}^{1*}) are generated by adjusting discipline-specific variables (X^1) to satisfy constraints

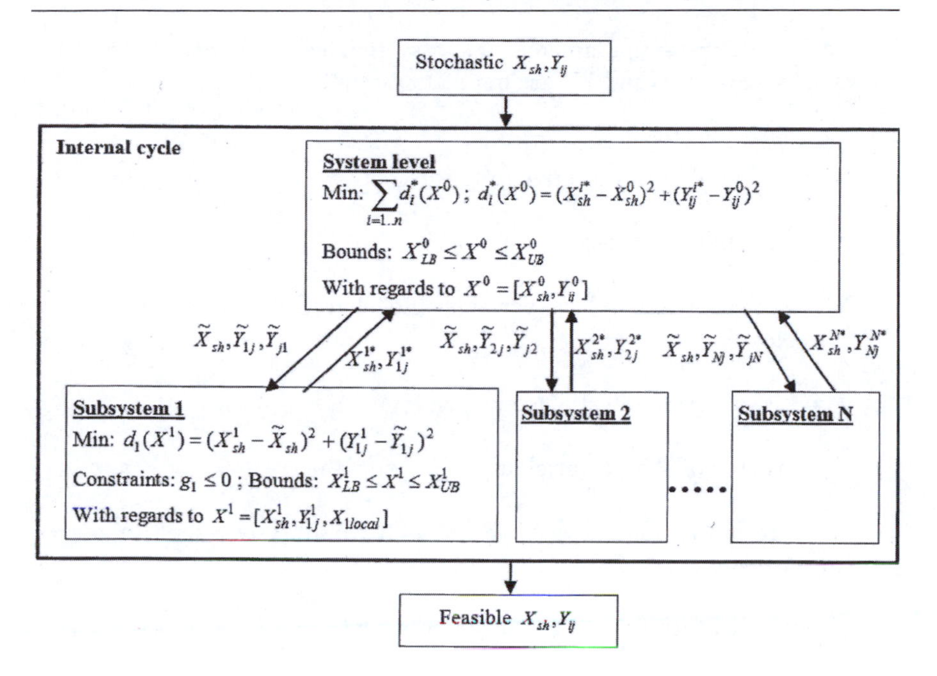

Figure 6.2 Formulation of the internal cycle.

at sub-system level (g_1). These optimal values are then sent to the system level where new values of interdisciplinary variables (X^0) are determined to reduce the compatibility function

$$\sum_{i=1...n} d_i^*(X^0)$$

The process is then restarted at sub-system level using the new target values $\tilde{X}^0$ sent from the system level. It is obvious that the system-level optimiser needs to call the sub-system optimisation process in each iteration. The iterations between the two levels are stopped when the change in the compatibility objective function at system level in two successive iterations is less than the predefined value.

The sub-system-level optimisers control their own local variables (X_{ilocal}) and manage the interdisciplinary variables ($X_{sh}^{\,i}$ and $Y_{ij}^{\,i}$). The system-level optimiser can also adjust the values of these interdisciplinary variables, which become known as auxiliary variables ($X_{sh}^{\,0}$ and $Y_{ij}^{\,0}$). The target values for the auxiliary variables assigned by the system-level optimiser are fixed during the process of sub-system-level optimisation, whereas the optimal values of these same variables ($X_{sh}^{\,i*}$ and $Y_{ij}^{\,i*}$) determined by the sub-system-level optimiser are fixed during the optimisation process at system level. Furthermore, both objective functions at system and sub-system level have the same formulation. The difference is that $X_{sh}^{\,i}$ and $Y_{ij}^{\,i}$ are variables and

$\tilde{X}_{sh}$ and $\tilde{Y}_{ij}$ are constant parameters at sub-system level, whereas $X_{sh}{}^{0}$ and $Y_{ij}{}^{0}$ are variables and $X_{sh}{}^{i*}$ and $Y_{ij}{}^{i*}$ are fixed at system level.

Where:

m = the number of objective functions

n = the number of disciplines in design

X^0 = vector of system level-variables (i.e. shared and coupling variables);

d_i = ith subsystem-level objective, $i \geq 1$;

X_{sh} = vector of shared variables;

Y_{ij} = vector of coupling variables, namely ith subsystem send value to jth sub-system;

g_i = vector of ith subsystem constraints, $i \geq 1$;

X^i = vector of ith subsystem variables, $i \geq 1$;

X_{ilocal} = vector of subsystem ith local variables, $i \geq 1$;

$(\bullet)^*$ = optimal value of variables;

$(\tilde{\bullet})$ = target value of interdisciplinary variables sent from system level to sub-system level.

External cycle of the PGACO framework

Although the internal cycle attempts to obtain compatible solutions, in some cases the compatible objective function cannot be minimised to zero or near to zero. Hence one aim of the external cycle enables these incompatible solutions to be excluded in the next generation in the course of Pareto-based GA. Furthermore in the external cycle new stochastic values of the interdisciplinary variables are generated through genetic operations, namely selection, crossover and mutation. The non-dominated sorting GA-II (NSGA-II) algorithm is used in the external cycle (Deb *et al.*, 2002). This algorithm is better than the simple Pareto-based GA (Fonseca and Fleming, 1993) because it introduces the parameter of crowding distance, which aims to obtain a uniform spread of solutions along a Pareto frontier (Deb *et al.*, 2002). The process of the external cycle is described below:

1 *Input interdisciplinary variables*
The process starts with selecting the values of interdisciplinary variables at random within their constraint bounds to create an initial population in the GA process. These are then used to start the internal cycle.
2 *Undertake the internal cycle* (see above for details)
3 *Check the interdisciplinary variables for compatibility between disciplines*
This stage identifies those interdisciplinary variables that have different values in different disciplines. If the compatible objective function ($d_1^* + d_2^*$) is close to zero, the interdisciplinary variables have equal values in all disciplines; the design objective functions (e.g. minimise capital cost) are then calculated. For the other interdisciplinary variables, the corresponding objective functions are assigned arbitrary large values in order for their related variables to be excluded from the next generation.
4 *Evaluate the objective functions values*
The objective functions in the external cycle are real design objectives, such as capital cost or annual running cost. Their values are assessed based on the Pareto rank and crowding distance. The design solutions with low Pareto rank and high crowding distance are regarded as the better ones. Every design solution is first assessed using the Pareto rank; for those design solutions with the same rank, the crowding distance is compared.
5 *Check for the terminal criterion in the external cycle*
The terminal criterion in this study is to complete a predefined number of generations. If this criterion is met, the external cycle will be terminated. Otherwise the external cycle goes to the next step, namely executing the genetic operators.
6 *Execute the genetic operators*
The genetic operators define the new individuals called offsprings for the next generation. It starts by selecting two individuals at random in the current generation and then choosing the one with the lowest Pareto rank or the highest crowding distance. This process is called a tournament and is repeated until a predefined number of individuals is obtained; these are the parent individuals. Finally the crossover and mutation are performed based on these parent individuals.
7 *Repeat steps 2 to 6 until the predefined number of generations is reached*
In each generation the offspring members created in the external cycle are sent to the internal one.

Application of the PGACO framework

This section explains how the PGACO framework is applied to the design of a typical three-storey office building. Data in this design scenario are collected from design documents and interviews with designers. It is therefore a realistic representation of a design task and is sufficiently complex to demonstrate both merits and limitations of the PGACO framework in multiobjective multidisciplinary building design. Wherever the required data were not available, appropriate assumptions were made.

Design case study

A south-facing room on an intermediate floor in a three-storey office block located in London is chosen as a design example; this is shown in Figure 6.3. A design structure matrix (DSM) is used to represent the links between the interdisciplinary variables that affect structural and heating, ventilation and air conditioning (HVAC) designs.

Design structure matrix representation

A number of variables and parameters contain interdisciplinary dependencies, which influence not only the final system design, but also the response from the discipline-specific objective function. The parameter-based DSM method is adopted here; this is a graphical tool that aids the designer in organising and structuring the design synthesis process, and shows the relationships between the various disciplines involved in a design problem (Yassine and Braha, 2003; Pektas and Pultar, 2006). There are three different forms of relationships between disciplines in the DSM method: parallel, sequential and coupled. These are illustrated in Figure 6.4.

The relationships between the 23 parameters for this design scenario are shown in Table 6.2. The parameters in the table represent the design sequence in the form of a square matrix. Dependencies between the

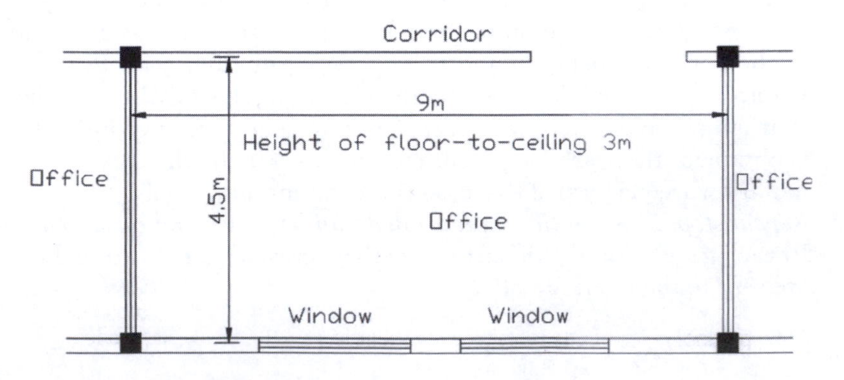

Figure 6.3 Office floor plan.

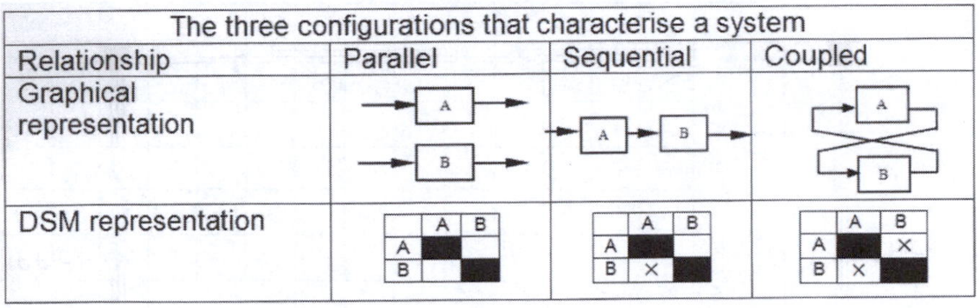

The three configurations that characterise a system			
Relationship	Parallel	Sequential	Coupled
Graphical representation	A B	A B	A B
DSM representation	A B A B	A B A B X	A B A X B X

Figure 6.4 The three configurations that characterise a system in the DSM representation (MIT DSM Research Group, 2005).

variables are indicated with the 'X' symbols on the off-diagonal cells. There are two differences in the way the DSM method is used in this case study:

- Variables are grouped by discipline; thus the first row and column represent the discipline name.
- Dependencies between the discipline-specific variables are not indicated.

Reading across a row shows the input variables; reading down a column shows the output variables. The interdisciplinary relationships between the variables are presented with the symbol 'X'. For instance the marks in row O of Table 6.2 denote that variable O (i.e. heating gain from floor, wall and roof) requires information from variables F, H, I and J (i.e. structural component material; floor, wall and roof dimension). If the design information is generated in the order A through to W and also processed in the same sequence, then the information required by each variable will have been generated by a predecessor task. It can be seen in the table that this is not always the case; for example, variable H requires information from variable R, as the position size of heating equipment affects the value of live load on a floor. In practice the information for variable R cannot be available before generating the value of variable H.

A further representation of the relationships between the interdisciplinary design variables that affect the architectural, structural and HVAC disciplines is shown in Figure 6.5. This presents the relationship between the architectural, structural and HVAC design for the case study and Figure 6.6 explains the meaning of the different information flow representations.

As Figure 6.5 shows, information is sent from the architectural design process to both the structural and HVAC systems simultaneously and includes space function, floor-to-ceiling height, orientation and glazing. In practice these variables are considered as fixed parameters at the start of the structural and HVAC design processes. However, if the structural and HVAC engineers cannot meet their own design criteria, they have to consult with the architect to change some of the parameters.

Table 6.2 Application parameter-based DSM in design scenario

Parameter		*Input*	*Architecture design variable*				*Structural design variable*									*HVAC design variable*									
Output			A	B	C	D	E	F	G	H	I	J	K	L	M	N	O	P	Q	R	S	T	U	V	W
Architecture design variable	Space function	A																							
	Orientation of building	B																							
	Height of floor to ceiling	C																							
	Window size and type	D																							
Structural design variable	Structural system type	E	X																						
	Structural component material	F																							
	Column layout	G																							
	Floor dimensions	H																		X	X				
	Wall dimensions	I																							
	Roof dimensions	J																							
	Cross-section of beam	K				X														X	X				
	Height of floor to floor	L			X																			X	
	Cross-section of column	M																							

HVAC Design Variable	Solar gain	N		X		X																			
	Heating gain from floor, wall and roof	O						X		X	X	X													
	Heating gain from occupant and equipment	P	X																						
	HVAC system type	Q																							
	Heating equipment size and position	R			X																				
	Cooling equipment size and position	S																							
	Air conditioning-beam integration scheme	T																							
	Air diffuse layout	U																							
	Air duct size	V																							
	Supply air temperature	W																							

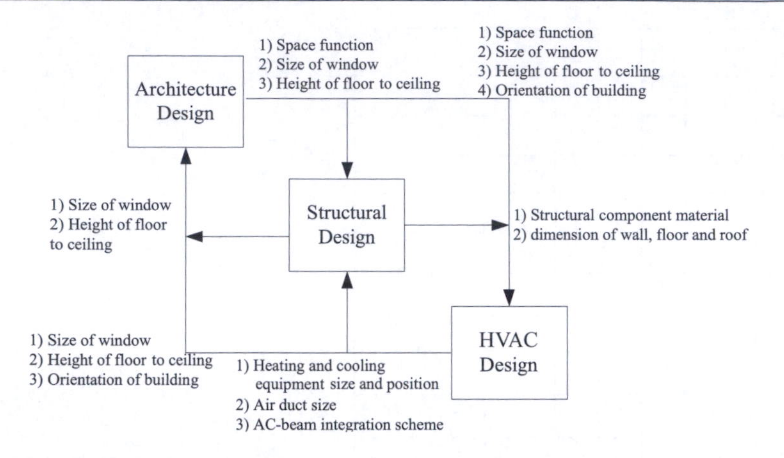

Figure 6.5 Interdisciplinary variables.

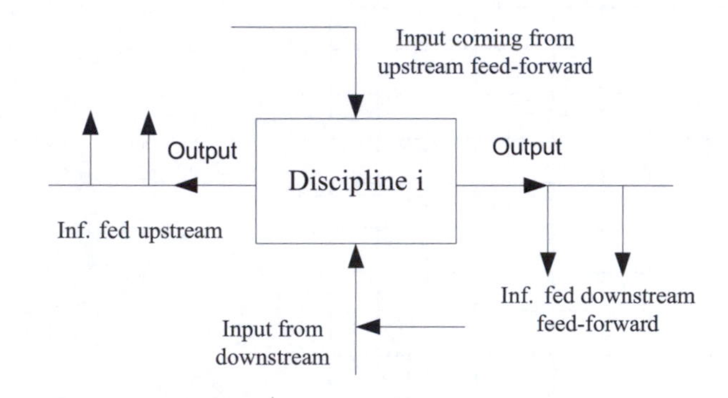

Figure 6.6 Definition of a module within the N-square diagram.

Figure 6.5 also illustrates the dependencies between structural and HVAC systems. The HVAC designer, for example, requires the materials and dimensions of the structural components to calculate the *U* values of wall, floor and roof. On the other hand, the structural designer requires information about the weight and position of heating and cooling equipments, which influence the live loads on the floor. Information related to the air ducts' sizes and their integration with the structural system is also needed by the structural designer and is a function of the floor-to-floor height.

Summary of information in the design case study

In order to simplify the design scenario, some variables are assigned fixed values before the start of the optimisation process; these include the orientation of building and floor-to-ceiling height. The case study includes two objective functions, six variables and 10 constraints; these are presented in Table 6.3. Table 6.4 includes the constructional and occupancy information.

Table 6.3 Objective functions, variables and constraints

Objective functions	*Variables*	*Constraints*
To minimise weight of column and beam F_1 (x_1, x_2, x_4, x_5)	Height of window (x_1) (mm)	Allowance for beam bending moment g_1 (x_1, x_2, x_3, x_4)
To minimise sum of peak cooling and heating load in summer and winter F_2 (x_1, x_2)	Depth of external wall (x_2) (mm)	Allowance for beam shear force g_2 (x_1, x_2, x_3, x_4)
	Depth of air duct (x_3) (mm)	Allowance for beam deflection g_3 (x_1, x_2, x_3, x_4)
	Depth of beam (x_4) (mm)	Allowance for column slenderness ratio g_4 (x_3, x_4, x_5)
	Depth of column (x_5) (mm)	Allowance for cross-section capacity of column with moments in-plan buckling g_5 (x_1, x_2, x_3, x_4, x_5)
	Supply air temperature in summer (x_6) (°C)	Allowance for cross-section capacity of column with moments out-of-plan buckling g_6 (x_1, x_2, x_3, x_4, x_5)
		Allowance for column axial force g_7 (x_1, x_2, x_3, x_4, x_5)
		Allowance for depth of external wall g_8 (x_2, x_5)
		Allowance for air change rate g_9 (x_1, x_2, x_6)
		Allowance for air velocity in the duct g_{10} (x_1, x_2, x_3, x_6)

Table 6.4 Constructional and occupancy details

Item	*Details*
External wall (opaque)	105 mm outer brickwork; x_2 mm inner brickwork; 13 mm lightweight plaster
Internal partition wall	13 mm lightweight plaster; 105 mm brickwork; 13 mm lightweight plaster
Internal floor/ceiling	50 mm screed, 150 mm dense cast concrete, 25 mm wood block; 16 mm plasterboard ceiling (density 391.2 kg/m^2, U = 1.5, Y = 2.9)
Window	Double glazed
Lighting	18.75 W/m^2 of floor area; in use 0900–1700 h
Occupancy	Occupied 0900–1700 h by six people, 80 W sensible heat output per person
Electrical equipment	Four computers of 150 W, in use 0900–1700 h
Mechanical ventilation	10.5 L/s fresh air per person

For the DSM representation, the design variables are expressed in terms of shared, coupling and local variables for the structural and HVAC disciplines; these are listed and explained in Table 6.5.

Table 6.5 Summary of all variables in the design scenario

Type of variable	*Variable*	*Comments*
Shared variable	Height of window (x_1)	In structural design, window size is a function of self-weight of exterior wall, which affects dead load to beam and column, whereas in thermal design it mainly affects heating loss in the room
Coupling variable (input of HVAC from output of structural design)	Depth of wall (x_2)	Depth of wall is decided during structural design, but calculation of *U* value of wall needs this variable in thermal design
Coupling variable (input of structural from output of HVAC design)	Depth of ventilation duct (x_3)	The services engineer will adjust depth of duct to satisfy noise requirement, the structural designer will calculate the height of floor-to-floor based on this variable
Structural local variable	Cross-section of beam and column (x_4, x_5)	This is a local variable in structural design. It is adjusted to satisfy structural strength, stability and so on
HVAC local variable	Supply air temperature in summer (x_6)	This is a local variable in the HVAC discipline to control supply air rate

The calculations are based on the following assumptions:

- The structural calculation is based on the combination of dead loads and live loads only. Wind loads and seismic loads are not considered.
- The calculations for the structural components are based on the elasticity analysis.
- The floor-to-floor height is the same in every storey.
- The density of steel is 7,850 kg/m^3.
- The dry resultant temperature in adjoining rooms is equal; hence heat flow occurs only through the outside window and wall.
- The office is located in the centre of London, so the window must be closed all day to avoid traffic noise. The air infiltration rate is equivalent to one air change per hour. The mechanical ventilation is based on a minimum fresh air requirement per person.
- With regard to the air conditioning–beam integration, this case study assumes the separation of service and structural zones; the details of this are shown in Figure 6.7.
- The thermal admittance (*Y*-value) does not change because, for a multilayered structure, the admittance is primarily determined by the characteristic of materials in the layer nearest to the internal surface.
- The peak heating load is assumed to happen at noon in January, and the peak cooling load occurs in July.

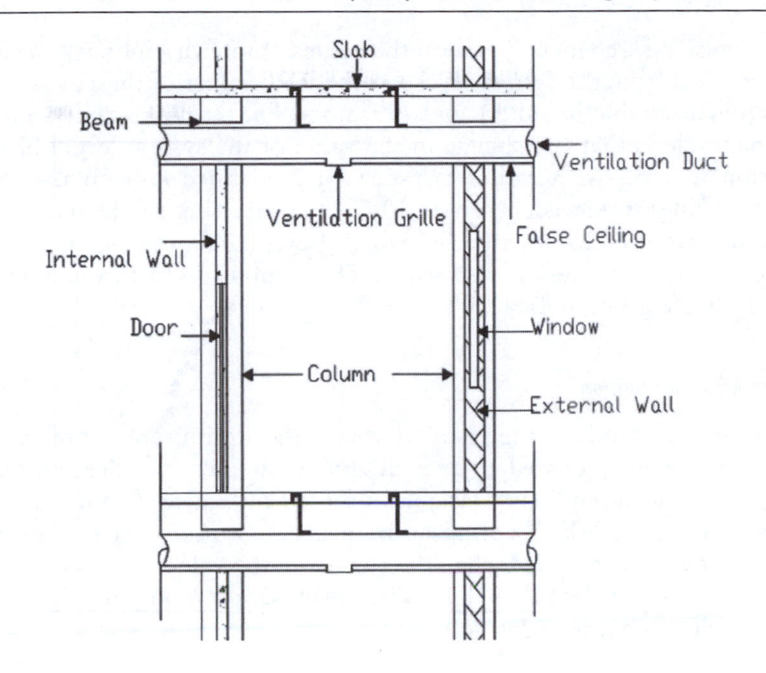

Figure 6.7 Details of the structural and services zones.

Application of the PGACO framework to the design scenario

As explained above, the PGACO framework includes two cycles, internal and external. The aim of the internal cycle is to obtain the feasible values of interdisciplinary variable (i.e. x_1, x_2 and x_3) through a two-level process, whereas the external cycle is used to implement the Pareto-based GA to optimise the objective functions F_1 and F_2. The detailed formulation of the two cycles within the context of the case study is presented as follows.

Internal cycle formulation

The design scenario can be decomposed in line with the two main disciplines (i.e. structural and HVAC designs). The design problem is represented at the system level of the internal cycle by a single objective function which is a the compatible function $d_1^* + d_2^*$. The problem is then decomposed into two sub-problems at the sub-system level. Sub-system 1 represents the structural system while sub-system 2 represents the HVAC system.

The system-level problem determines the value of interdisciplinary variables, namely shared variables (x_1) and coupling variables (x_2 and x_3) to optimise the objective function $d_1^* + d_2^*$. The sub-system processes aim to minimise discrepancies between the values of interdisciplinary variables passed down from the system-level and the local values of these, while also satisfying discipline-based design constraints. The terminal criterion of the internal cycle is that the change in the value of the system-level objective function in successive iterations is less than a predefined value. In this study, this predefined value is chosen as 10^{-5}. The following mathematical formulations are used to describe the internal cycle processes of the PGACO framework for this design case study. The symbols used throughout this case study are given in Table 6.3.

System-level formulation

At the system level of the internal cycle, the optimisation process co-ordinates the data related to two design disciplines by selecting values for all the interdisciplinary variables used to minimise $d_1^* + d_2^*$. In each iteration, the system-level optimiser initiates the two sub-system processes. Such iterations are stopped when the change in the value of $d_1^* + d_2^*$ in two consecutive runs is less than 10^{-5}. The optimisation model for the system level problem is expressed as follows.

SYSTEM-LEVEL MODEL

Minimise: $d_1^* + d_2^*$

$$d_1^* = (x_1^{1*} - x_1^0)^2 + (x_2^{1*} - x_2^0)^2$$

$$d_2^* = (x_1^{2*} - x_1^0)^2 + (x_3^{2*} - x_3^0)^2$$

With respect to: x_1^0, x_2^0, x_3^0

Constraints: $1000\,\text{mm} \le x_1^0 \le 3000\,\text{mm}$; $0\,\text{mm} \le x_2^0 \le 120\,\text{mm}$; $100\,\text{mm} \le x_3^0 \le 700\,\text{mm}$

x_1^{1*}, x_2^{1*} and x_1^{2*}, x_3^{2*} are optimal values for sub-systems 1 and 2 respectively. During the system-level optimisation these are fixed while the interdisciplinary variables x_1^0, x_2^0, x_3^0 are adjusted to minimise the function $d_1^* + d_2^*$.

Sub-system-level formulation

The structural designer determines the dimension of the structural components (e.g. floor, wall) to effectively support all the loads imposed on the building; the building services engineer must achieve occupants' indoor comfort by way of designing the building envelope and the mechanical services (e.g. heating and air conditioning systems). In this case study, the dimensions of beams and columns (x_4, x_5) and the air supply temperature in the duct in summer (x_6) are local variables within the structural and HVAC sub-systems respectively. Within the PGACO framework, the sub-system optimisers can also control the interdisciplinary variables (x_1, x_2, x_3). They receive the target values of the windows dimensions, thickness of walls and size of ducts from the system level, then aim to obtain values as close to these as possible by varying the discipline-specific variables while remaining within the limits of the local constraints.

The mathematical formulations for the structural and HVAC sub-systems are expressed as follows:

SUB-SYSTEM 1: STRUCTURAL SUB-SYSTEM FORMULATION

Minimise: $d_1 = (x_1^1 - \tilde{x}_1)^2 + (x_2^1 - \tilde{x}_2)^2$

With respect to: x_1, x_2, x_4, x_5

Constraints:

Allowance for beam bend moment: $g_1\ (x_1^{\,1}, x_2^{\,1}, \tilde{x}_3, x_4)$;

Allowance for beam shear force: $g_2\ (x_1^{\,1}, x_2^{\,1}, \tilde{x}_3, x_4)$;

Allowance for beam deflection: $g_3\ (x_1^{\,1}, x_2^{\,1}, \tilde{x}_3, x_4)$;

Allowance for column slenderness ratio: $g_4\ (\tilde{x}_3, x_4, x_5)$;

Allowance for cross-section capacity of column with moments in-plan buckling: $g_5\ (x_1^{\,1}, x_2^{\,1}, \tilde{x}_3, x_4, x_5)$;

Allowance for cross-section capacity of column with moments out-of-plan buckling: $g_6\ (x_1^{\,1}, x_2^{\,1}, \tilde{x}_3, x_4, x_5)$;

Allowance for column axial force: $g_7\ (x_1^{\,1}, x_2^{\,1}, \tilde{x}_3, x_4, x_5)$;

Allowance for the depth of external wall: $g_8\ (x_2^{\,1}, x_5)$;

Bounds: $1000\,\text{mm} \le x_1^{\,1} \le 3000\,\text{mm}$; $0\,\text{mm} \le x_2^{\,1} \le 120\,\text{mm}$;
$200\,\text{mm} \le x_4 \le 1200\,\text{mm}$; $100\,\text{mm} \le x_5 \le 400\,\text{mm}$

SUB-SYSTEM 2: HVAC SUB-SYSTEM FORMULATION

Minimise: $d_2 = (x_1^{\,2} - \tilde{x}_1)^2 + (x_3^{\,2} - \tilde{x}_3)^2$

With respect to: $x_1^{\,2}$, $x_3^{\,2}$, x_6

Constraints:

Allowance for air change rate: $4 \leq g_9\ (x_1^2, \tilde{x}_2, x_6) \leq 20$;

Allowance for air duct velocity: $2 \leq g_{10}\ (x_1^2, \tilde{x}_2, x_3^2, x_6) \leq 4.5$;

Bounds: $1000\,\text{mm} \leq x_1^2 \leq 3000\,\text{mm}$; $100\,\text{mm} \leq x_3^2 \leq 700\,\text{mm}$; $15°\text{C} \leq x_6 \leq 20°\text{C}$;

$\tilde{x}_1$, $\tilde{x}_2$ and $\tilde{x}_3$ are target values sent from the system level and fixed at the sub-system level.

External cycle formulation

The external cycle of the PGACO framework aims to optimise the original design objectives. In this case study, two design objective functions are minimised: the first one is total weight of beam and column; the second is the sum of peak heating load in winter and cooling load in summer. Some assumptions are made for the two objective functions. With regard to the first objective, the cross-section of all beams is the same and so are all columns; the calculation of this objective function is based on the weight of one beam and one column. The second objective function is calculated on the assumption that the occurrence of the hottest day in London is in July.

The mathematical formulation of the external cycle is as follows:

Minimise: F_1 and F_2

F_1 is the total weight of column and beam:

$$F_1 = [(18000 + 18x_4) \times 9000 + (3600 + 8x_5) \times (3000 + x_3 + x_4 + 100)] \times 7850 \times 10^{-9}$$

F_2 is the total peaking cooling load and heating load:

$$F_2 = Q_k (x_1, x_2) + Q_f (x_1, x_2)$$

With respect to: x_1, x_2, x_3

Bound: $1000\,\text{mm} \leq x_1 \leq 3000\,\text{mm}$; $0\,\text{mm} \leq x_2 \leq 120\,\text{mm}$; $100\,\text{mm} \leq x_3 \leq 700\,\text{mm}$

The optimiser in the external cycle only manages the interdisciplinary variables (x_1, x_2, x_3), while the local variables (x_4, x_5) are fixed during the course of the optimisation. In addition, there are no constraints in the external cycle optimisation. This characteristic is favoured in the use of the Pareto-based GA.

The PGACO optimisation process

Figure 6.8 shows the flow chart for the internal and external cycle, as well as inputs and outputs between the system and sub-system levels in the internal cycle.

This flow chart starts in the external cycle, where the values of interdisciplinary variables (i.e. x_1, x_2, x_3) are initialised and sent to two sub-systems in the internal cycle. After completing the optimisation run within the sub-system level, the optimal values of interdisciplinary variables are sent to the system level ($x_1^{1*}, x_2^{1*}, x_1^{2*}, x_3^{2*}$), the system-level optimiser sets new values for the interdisciplinary variables (x_1^0, x_2^0, x_3^0) to minimise the objective function $d_1^* + d_2^*$, and then sends these new target values (i.e. $\tilde{x}_1$, $\tilde{x}_2$, $\tilde{x}_3$) back to the two sub-systems to restart the optimisation run. Such iterations between system and sub-system levels are stopped when the change in the objective function $d_1^* + d_2^*$ in two successive iterations is less than 10^{-5}. Finally the value of $d_1^* + d_2^*$ with the corresponding interdisciplinary variables is sent out of the internal cycle, meaning that the internal cycle has been completed.

The first step of the external cycle is to evaluate whether the design solutions are compatible using the system-level objective function $d_1^* + d_2^*$. 'Compatibility' here means that the same interdisciplinary variables receive equal values in the two sub-systems. If $d_1^* + d_2^* \leq 0.01$, it implies that it is a compatible design, and then the external cycle's objective functions are calculated (F_1 and F_2); otherwise both F_1 and F_2 are assigned a value of 10^6. After that, F_1 and F_2 are assessed using the principles of Pareto rank and crowding distance (Deb *et al.*, 2002). Then the terminal criterion of the external cycle is evaluated, namely the predefined number of iteration in this case study. If this criterion is not met, the genetic operations including selection, crossover and mutation are performed. This step is to generate new stochastic values for the interdisciplinary variables to run the internal cycle again.

Implementation setup

The Sequential Quadratic Programming (SQP) optimisation toolkit in Matlab was used to perform both the sub-system- and system-level optimisation in the internal cycle. The non-dominated sorting GA-II (NSGA-II) was adopted for the external cycle optimisation. The algorithm was written in Matlab as a separate program. In this program the simulated binary crossover and polynomial mutation was used; the crossover probability is below 90 per cent and mutation probability is below $1/n$, where n is the number of decision variables for real-code GAs. The distribution indices for crossover and mutation are 20 and 20 respectively (Deb *et al.*, 2002). Hence, for this study, the parameters include:

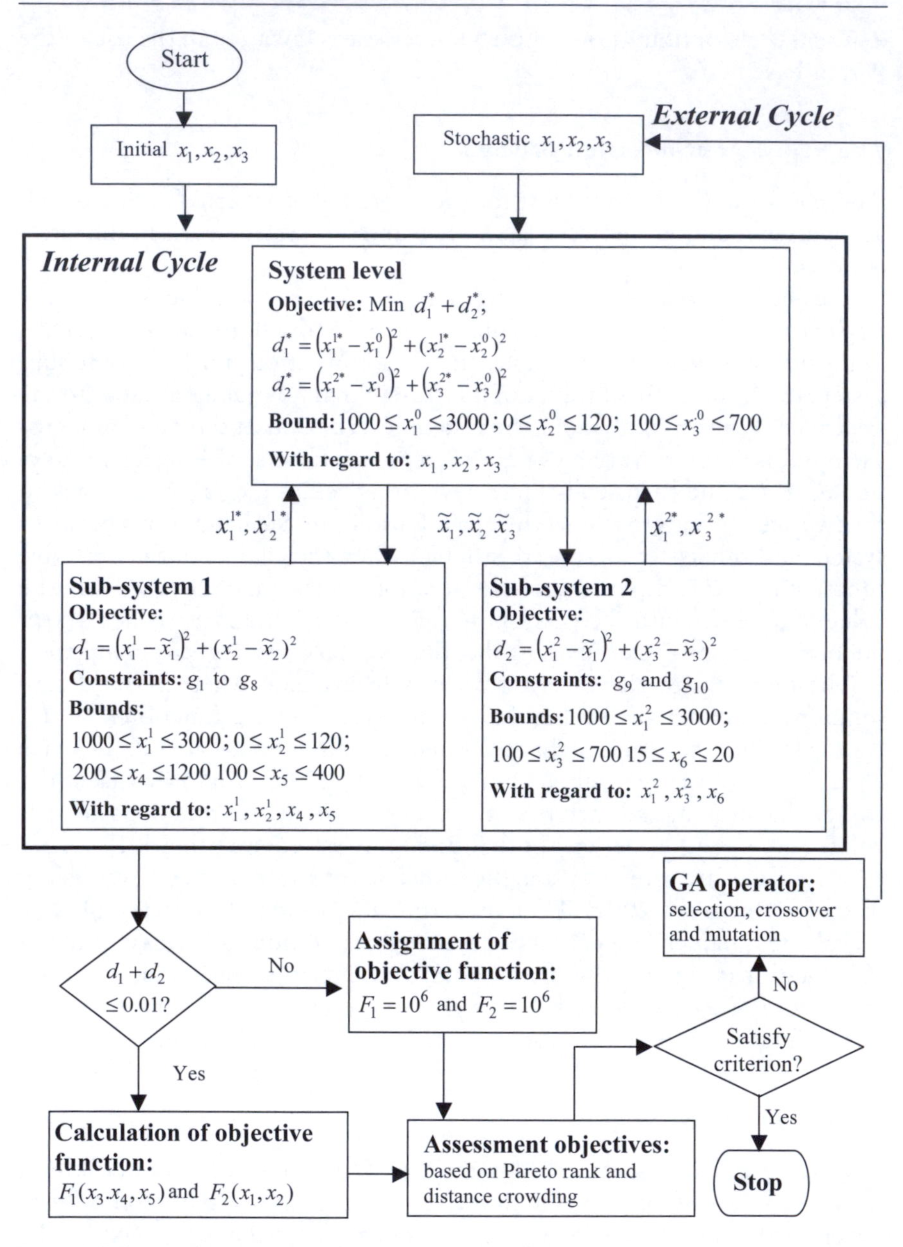

Figure 6.8 Representation of the design scenario in the PGACO framework.

- number of generations = 60;
- population size = 60;
- crossover probability < 0.9;
- mutation probability < 0.25.

All design variables are considered to be continuous.

The PGACO results

In this section, the results of the design case study optimised through the PGACO formulation are analysed, based on the results from the 60 iterations and the coordination process within the internal cycle.

Analysis of the results

After completing 60 iterations in the external cycle, the Pareto frontier shown in Figure 6.9 is generated. This curve of trade-off between the two objective functions (F_1 and F_2) indicates that a further decrease in F_1 will be at the expense of increasing F_2. The six representative design solutions described in Table 6.6 are selected from the collective design solutions based on the maximal or minimal values of the objective functions or variables.

The 60 sets of variables including the two objective functions and three interdisciplinary variables are illustrated in Figure 6.10. This shows that the window size curve (x_1) has a similar variation pattern to the second objective curve (F_2) and an inverse shape to the first objective curve (F_1). This suggests that increasing the value of x_1 will increase the value of F_2 but decrease the value of F_1. In addition, although the depth of wall (x_2) is also the function of F_1 and F_2, the figure shows that changes in the function of F_1 and F_2 do not follow changes in this variable. It implies that the functions of F_1 and F_2 are more sensitive to x_1 than to x_2. This is why the design solution

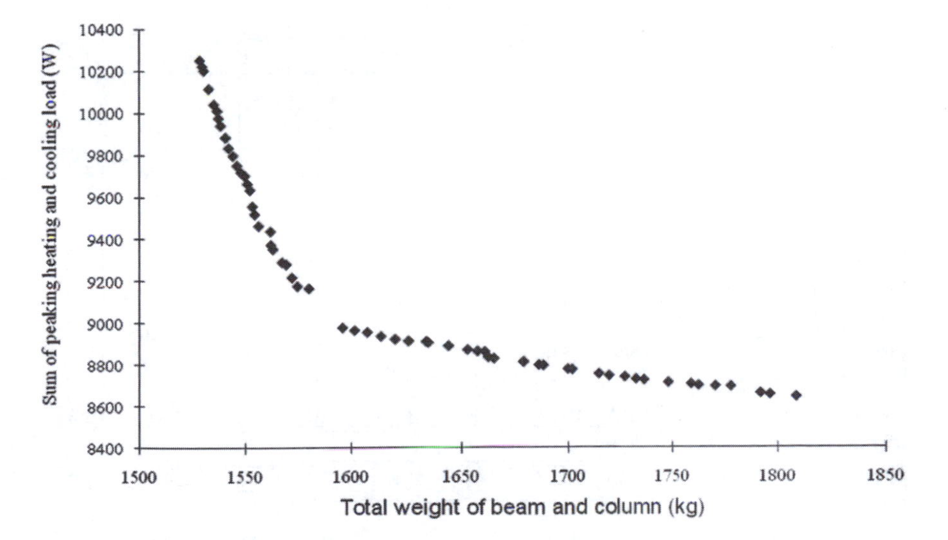

Figure 6.9 PGACO solutions for the design scenario.

in which F_1 has a minimum value and F_2 has a maximum value corresponds to where x_1 is at the maximum (e.g. solution no. 1 in Table 6.6), whereas the design solution in which x_2 is at the minimum may not correspond to F_1 being at the minimum and F_2 being at the maximum (e.g. solution no. 6 in the table).

Note: values of variables and objective functions in Figure 6.10 are reduced on certain scales in order that they could be presented as a drawing.

Analysis of the coordination within the internal cycle

Because not all results from the internal cycle are feasible design solutions, trade-offs are required in the internal cycle. The convergence process is shown through the evolution in the values of x_1, x_2 and x_3 at system level. These start as 2000 mm, 112 mm and 600 mm respectively and become 1075 mm, 0 mm and 562 mm respectively when the value of the system-level objective function ($d_1 + d_2$) reaches 1.51×10^{-10} after the completion of the internal cycle. Such a solution could be regarded as an acceptable design.

Figure 6.11 shows the discipline-specific convergence histories for three system-level iterations with respect to the variable x_1. In the first iteration, the system level allocates x_1 the value of 3000 mm. After checking all of its discipline-specific constraints, the structural sub-system accepts this value while the HVAC sub-system responds with a request to reduce this value to 1955 mm. Based on the optimal solution provided by each sub-system, a system-level step is taken to alter the value of x_1 to 914 mm. Both sub-systems then request increasing this new value to 1000 mm after conducting their

Table 6.6 Six representative solutions of the case study

Number of solution	F_1 *(kg)*	F_2 *(W)*	x_1 *(mm)*	x_2 *(mm)*	x_3 *(mm)*	x_4 *(mm)*	x_5 *(mm)*	x_6 *(°C)*
No. 1 (min F_1, max F_2, max x_1, min x_2, min x_4, min x_5)	1529	10251	1812	0	389	731	220	15.2
No. 2 (max F_1, min F_2, min x_1, max x_2, max x_4, max x_5)	1808	8647	1000	120	394	791	242	15.1
No. 3 (min x_3)	1556	9466	1311	0.3	354	740	221	15.2
No. 4 (max x_3)	1661	8865	1000	31.3	501	759	230	16.6
No. 5 (min x_6)	1550	9704	1464	0.8	371	736	221	15.0
No. 6 (max x_6)	1596	8981	1000	0	454	744	226	16.7

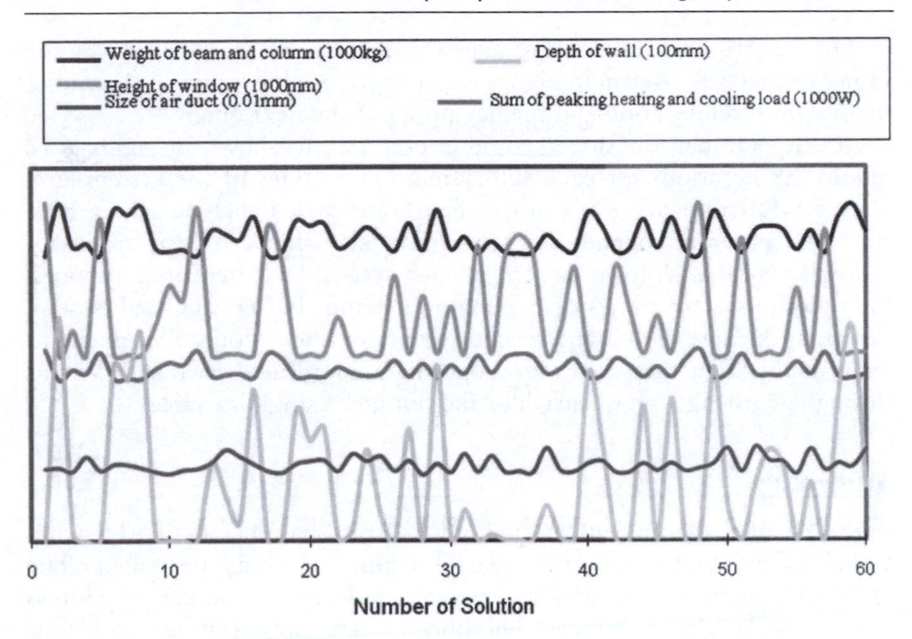

Figure 6.10 Pattern of the 60 solutions.

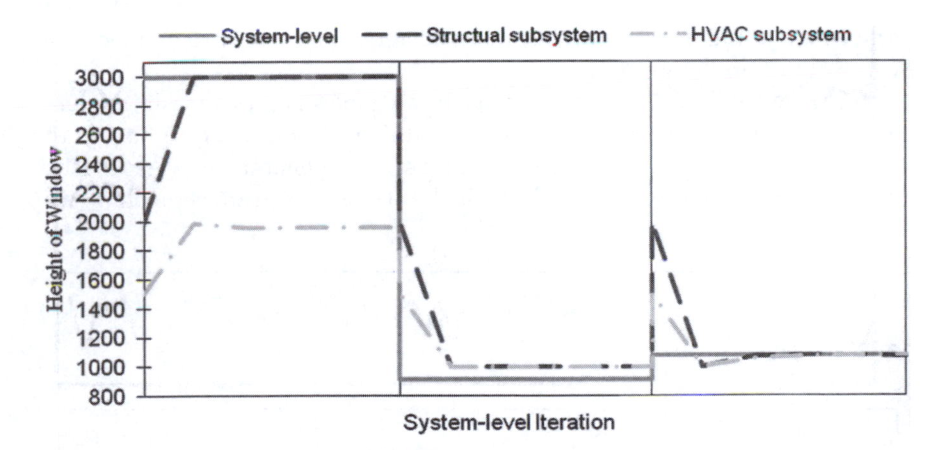

Figure 6.11 System-level coordination of value of window's height.

own optimisation runs. In the third iteration, the system level increases the value of x_1 to 1075 mm. Taking this new value, both sub-systems are able to remain within the constraints and hence agree on this value of x_1. Through this repeated process, the system level ensures the interdisciplinary compatibility of the design solutions. On the other hand some fixed initial values are assigned to the sub-systems; for example 2000 mm and 1500 mm are allocated as initial values for the window's height in each of the iterations within the structural and HVAC sub-systems respectively. This is why the values of each sub-system in Figure 6.11 experience big jumps at the end of the system-level iteration and the beginning of the next one.

In this example, the three system-level iterations shown in Figure 6.12 require 18 iterations for each sub-system. The values of the system-level objective function ($d_1^* + d_2^*$) in this figure are scaled such that they have the same order of magnitude. This figure also illustrates the trends of movement of $d_1 + d_2$ towards zero in each system-level iteration, although the two sub-system values of x_1 remain unchanged after a few sub-system iterations during the first and second system-level iterations. This phenomenon implies that the system-level optimiser coordinates two sub-systems through adjusting a set of variables and not just a single variable.

Summary

In this chapter, the implementation of the Pareto-based Genetic Algorithm Collaborative Optimisation (PGACO) framework was described. The PGACO framework provides a two-cycle multilevel structure to address large-scale, multidisciplinary building design optimisation problems. Local, shared and coupling variables are optimised using a Pareto optimality approach for multiobjective searches which contain multiple solutions rather than a single optimum; GA was also adopted in order to enhance the search ability in design space.

In the internal cycle, a set of stochastic variables is set to feasible values in a two-level process. The external cycle undertakes the Pareto-based GA to solve the multiobjective design problem and generate new stochastic values of interdisciplinary variables. Such a two-cycle framework has two

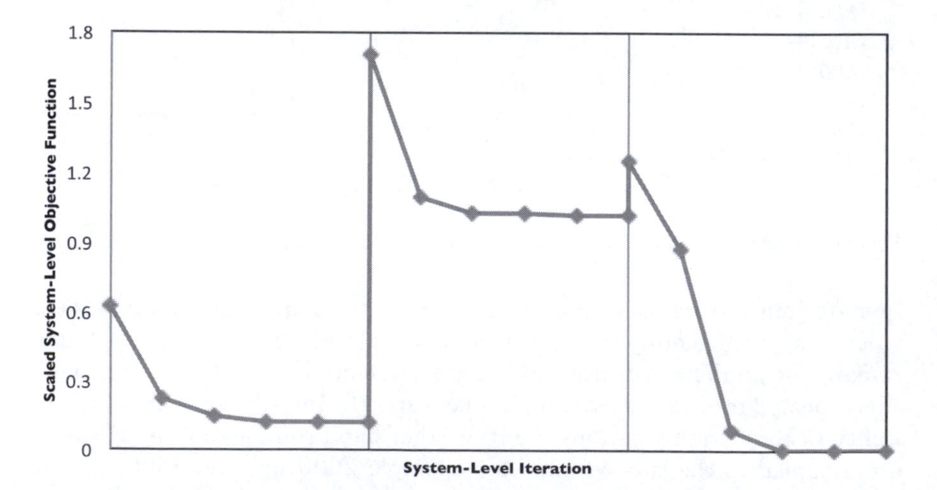

Figure 6.12 Progressions in system-level objective function.

distinguishing features. On the one hand, the Pareto-based GA is easy to implement because the optimisation problem of the external cycle is non-constrained. On the other hand, the framework avoids the delinquent nature of the MOCO formulation because constraint functions at the system level of the MOCO formulation are added up to constitute the single objective function at system level within the internal cycle.

Finally the PGACO framework is applied to a multiobjective multidisciplinary design scenario. It gives a step-by-step explanation of the main stages of the process including the decomposition of this design scenario according to the disciplines involved, the formulation of the problem based on the PGACO framework, and the implementing of the optimisation process. The design structure matrix is used in the case study to represent the interdependencies between the disciplinary design variables. In the case study, structural and HVAC disciplines were included whereby the optimisation aimed to minimise the total weight of the structural frame and the sum of peak heating and cooling load.

Chapter 7

Change management

Bilge Erdogan, Dino Bouchlaghem and Yasemin Nielsen

Introduction

Construction organisations are always faced with the need to cope with changes during projects; most of these are design related and, in the vast majority of cases, inevitable even after the most detailed design studies. Besides handling changes at project level, construction companies are also required to consider their impact at organisational level. This is usually more challenging for construction companies which have geographically dispersed organisational structures, are multidisciplinary in nature, and manage one-off projects with different participation in each. However, in an environment characterised by ever-increasing customer expectations in a global market, change management can play a key role in the quest by organisations to achieve competitive advantage.

Changes in projects have many triggers and are unavoidable; however, their impact can be minimised through a well-managed change strategy. The impacts of changes on the project, organisation and people vary according to their type and nature, and most importantly according to how they are managed. Managing change is important to maximise value, minimise risks and ensure that these are shared equitably and fairly.

This chapter reviews change and change management concepts in construction at both project and organisational levels. The categorisation and nature of changes are reviewed; enablers and barriers for managing the changes are discussed together with theories and tools from the literature. Particular attention is given to organisational strategic change management as this is an important step for the effective implementation of collaboration environments in construction.

Project change management

Project changes are considered to be any additions, deletions or revisions to the project aims and scope, and can affect the project cost and schedule (Ibbs *et al.*, 2001). Lazarus and Clifton (2001) define change in a construction project as anything that impacts:

- the scope, requirements or brief for the project;
- the capital cost or whole-life cost or value of the project;
- the time required to design or construct the project;
- the project team relationships and appointments;
- project-associated risk allocation or scope; and
- the form of procurement.

Classification of changes in project management

Different classifications of project changes exist and are mostly based on their impact, necessity and timing. According to their impact, the Construction Industry Institute (CII, 1994) considers changes to be either *beneficial* – those that can help to reduce cost, schedule or a degree of difficulty and are to be encouraged – or *detrimental* – those that reduce owner value or have other negative impacts on the project.

Another classification proposed by CII (1994) is based on their necessity: *required changes* are necessary to meet the basic venture objectives, regulatory/legal requirements or defined safety and engineering standards; *elective changes* are proposed to enhance the project, but are not required to meet the original project objectives.

Other classifications are based on their timing; for example, a *pre-fixity change* occurs during design development whereas a *post-fixity* change is usually encountered after (Lazarus and Clifton, 2001). The post-fixity changes are also divided into two types depending on their urgency: the urgent type are those that concern the design work agreed with the client and that require completion within six weeks or less; the non-urgent post-fixity changes are those not required to be completed within six weeks. Further classifications include *anticipated and emergent changes*, *proactive and reactive changes*. These classifications are summarised in Table 7.1.

Nature of changes in project management

Changes that affect project management are commonly grouped into four main types: changes in distant project environment, changes in close environment, project changes and key integrative process changes:

1. *Distant project environment changes* are caused by external uncertainties and include those resulting from factors related to political, legal, normative, social, economic, financial, ecological, technological and organisational aspects and other external factors influencing the project.
2. *Close environment changes* are those caused by internal uncertainties and are concerned with external relations within the parent organisation including target market of products, services and solvent demand;

Table 7.1 Classification of project changes

Criterion	*Project changes*	
Type of impact	Beneficial changes: reduce cost, schedule or degree of difficulty	Detrimental changes: reduce owner value, have negative impact on the project
Need for change	Required changes: implemented to meet the objectives or regulatory/legal/safety/ engineering requirements/standards	Elective changes: enhance the project, but are not required to meet the original objectives
Initiation nature/ responsiveness of change	Emergent/reactive changes: unplanned, unexpected; the response is after the occurrence	Anticipated/proactive changes: expected before they occur, therefore necessary actions are taken

Sources: CII (1994); Burnes (1996).

the concept of strategic organisational development and its policies; organisation forms and structures; production systems and technologies; organisation internal infrastructure; company's behaviour, culture and system; communication methods between local companies and international business community.

3 *Project changes* are those that can occur as a consequence of internal and external drivers and include changes in scope, quality, time, cost, risk, contract/procurement, human resources, and communications.
4 *Key integrative process changes* are related to the processes and procedure for providing appropriate actions for the above three types and include changes in project planning, execution, control systems and documentation.

The sources of change at project level are summarised in Table 7.2.

Causes and results of changes

Changes in projects are usually caused by variations (change orders); the need for rework (due to design changes and design and construction errors); and consequences of unexpected external events such as industrial action and inclement weather.

Design changes, also referred to as engineering changes, are those related to modifications in forms, fits, functions, materials and dimensions of products and constituent parts. Engineering changes constitute some of the most significant problems in both the construction and the manufacturing industries. There are three types of engineering changes that depend on when they occur in the design process:

Table 7.2 Sources of changes at project level

External reasons	*Internal reasons*
Changes regarding economic and financial issues	Changes in the organisational culture
Changes in environmental issues	Changes in the system of project planning
Changes in ecological issues	Changes in the project plan execution
Technology changes	Changes in the overall change control system
Changes in the standards and regulations	Changes in the documentation system
Political changes	Ineffective decision making
Force majeure	Design improvements
	Unexpected weather conditions
	Design error
	Designer change of mind
	Changed design parameters
	Contract disputes
	Changes in the project

Sources: Kast and Rosenweig (1974); Kitchen and Daly (2002); Lazarus and Clifton (2001).

- engineering changes during concept design; their impact is not considered to be significant;
- engineering changes after concept design; these tend to cause greater disruption as the production is usually under way;
- engineering changes during major reconstruction.

In a study carried out by Cox *et al.* (1999) examining the historical data from change order request procedures in construction, it was found that, in monetary terms alone, the direct cost of post-contract design changes amounts to between 5.1 per cent and 7.6 per cent of the total project cost. This highlights the importance of change order management in construction projects. The most common causes of change orders in construction can be summarised as (Cox *et al.*, 1999; Love *et al.*, 2002; Hsieh *et al.*, 2004):

- changes in client requirements;
- design errors; and
- unforeseen circumstances regarding site conditions or administrative issues such as change of work rules/regulations, change of decision-making authority, special needs for project commissioning, and ownership transfer.

The majority of change orders are due to design errors related to mistakes in quantity estimates, inadequate planning and arrangement of contract interfaces, inconsistencies between drawings and site conditions, and inadequate specifications. Concurrent engineering and design and build contracts are believed to reduce the number of engineering changes and offer better coping mechanisms during the construction stage provided that there are good communication systems and a focus on the customers' needs (Faniran *et al.*, 2001; Lau *et al.*, 2003). The requirements for a successful collaborative engineering change management are communication support, involvement of all relevant parties, working toward a consensus, control of the process, and knowledge of the scope of impact (Rouibah and Caskey, 2003).

Changes can have direct or indirect effects on the project team (Lazarus and Clifton, 2001). Direct effects can result in changes in project information, outputs and communication procedures; additional time and cost; reorganisation and rescheduling of work; changes in production schedules and deliveries; and the introduction of acceleration measures to maintain the project programme. Potential indirect effects include increased coordination failures and errors; increased waste in the process from abortive work and out-of-sequence working; reduction in productivity, quality of the product and level of profit; uncertainty; and lower morale. In addition, changes in projects always have consequences such as breaking of project momentum, increased overhead and equipment costs, scheduling conflicts, rework, and decreased labour efficiency (Sun *et al.*, 2006).

Managing changes in construction projects

Many previous studies provide guidelines on how to manage change. The Construction Industry Institute (CII, 1994) defined the project elements that need to be considered as part of the change management process as Project Scope, Project Organisation, Work Execution Methods, Contracts and Risk Allocation. The principles of effective change management according to the CII are:

- promote a balanced change culture;
- recognise change;
- evaluate change;
- implement change; and
- continuously improve from lessons learned.

A change management framework enabling users to produce a rich description of the change event was proposed by Sun and colleagues (2006) and consisted of four main parts:

1 a Change Dependency Framework, which provides a hierarchical structure with four levels: the first level consists of the key activities of a generic change management model and the other three are decompositions giving increased levels of detail;
2 a Change Prediction Tool, which aims to predict the changes in the construction project and links together the change, the project

characteristics, causes of the change and the impact of the change through a fuzzy logic approach;
3 a Workflow Tool, a software system created to identify the workflow changes by matching two versions of a workflow specification;
4 a Knowledge Management Guide, which explores the role of knowledge in managing project change in collaborative team settings.

Love and colleagues (2002) adopted a systems approach and suggested that the dynamics of a project should be evaluated and monitored using the following functions:

- planning for being proactive to change;
- organising: allocating tasks to people, requesting resources and coordinating all tasks into a working system;
- commanding: leading, delegating, communicating, motivating, coordinating, cooperating with and disciplining people;
- controlling: establishing standards and methods for measuring performance and determining the deviations from the planned requirements in terms of cost, time, quality and safety.

A parameter-based approach to engineering change management to support multicompany concurrent engineering efforts through the facilitation of information exchange, retrieval, sharing and use was proposed by Rouibah and Caskey (2003), and Huang and colleagues (2001) introduced a web-based system for engineering change management and discussed its design, development and implementation in the manufacturing industry. Although this targets the manufacturing industry, the basic principles are also applicable to construction.

Organisational change management

Organisational changes affect processes, functions, values, beliefs, human behaviour, power distribution and the way organisational processes are managed (Cao *et al.*, 2000). These are all interconnected and can affect one another.

Classification of organisational change

There are different classifications of organisational change; these are based on various criteria:

- According to their strategic significance, changes can be strategic or non-strategic. *Strategic change* refers to non-routine, non-incremental and discontinuous change which alters the overall orientation of the organisation and/or components of the organisation (Tichy, 1983).

Changes which do not affect the overall orientation of the company and do not result in a drastic difference are termed *non-strategic change*.

- Based on the speed of the transformation in the organisation, changes can be *incremental* or ongoing change that is routinely necessary for any organisation to adapt to its environment, or *radical*, referring to the sort of change that necessitates a thorough re-examination of all facets of an organisation (Cao *et al.*, 2000). Incremental change is also known as *gradual change* while radical change is sometimes referred to as *quantum change*. Cummings and Worley (2005) consider incremental change as fine-tuning the organisation whereas quantum changes entail fundamentally altering how an organisation operates.
- According to how they are initiated, they can be *emergent change*, driven from the bottom up in an open-ended continuous process of adaptation to changing conditions and modifications; *planned change*, which arises from a rational and systematic analysis of the social and organisational problems in question (Burnes, 1996); or *anticipated change*, which is not planned by the organisation but its occurrence is expected.
- Based on the way the change is implemented in the organisation, there are three categories: *developmental change* follows a continuous process; *transitional change* includes a period for transition between the old and new states of the organisation; and *transformational change* may involve both developmental and transitional changes.

The different classifications of organisational changes in the literature are summarised in Table 7.3.

Depending on the type of change, an organisation can use different approaches for its management (Figure 7.1). Incremental change triggered by the anticipation of future events needs a process called tuning. A strategic change driven by the anticipation of external events requires reorientation. Adaptation is used to respond to incremental changes carried out as a reaction to external events whereas reactive strategic changes necessitate recreation.

When considering the organisation's state before and after the change, three different approaches to the rate of change were identified (Figure 7.2). Approach A shows a directed change carried out at a specified pace and intention. Although this approach can have some competitive advantages, if conditions change, the time to gather information and feedback for the integration of directed and non-directed processes might not be adequate. Approach B also shows an intentional change that includes some planned built-in stops during the process to analyse the conditions and integrate some non-directed elements during the transition. Approach C deals with both directed and non-directed change processes through looping cycles of continuous feedback, monitoring and assessment, and is believed to bring more positive long-term results than the other two approaches (Felkins *et al.*, 1993).

Table 7.3 Classification of organisational changes

Criteria	*Organisational changes*
The difference in the organisation due to change (Tichy, 1983)	Strategic changes Non-routine, non-incremental and discontinuous, alter the overall orientation of the organisation
	Non-strategic changes Do not affect the overall orientation of the company, do not result in a drastic difference
Speed of the transformation in the organisation (Cao *et al.*, 2000; Cummings and Worley, 2005)	Incremental/gradual Routinely necessary for any organisation to adapt to its environment
	Radical/quantum changes Necessitate a thoroughgoing re-examination of all facets of an organisation
Initiation nature (Burnes, 1996)	Emergent changes Driven from bottom up; an open-ended and continuous process of adaptation to changing conditions
	Planned changes Result of action research and an analysis of the social and organisational problems in question
	Anticipated changes Not planned by the organisation but expected

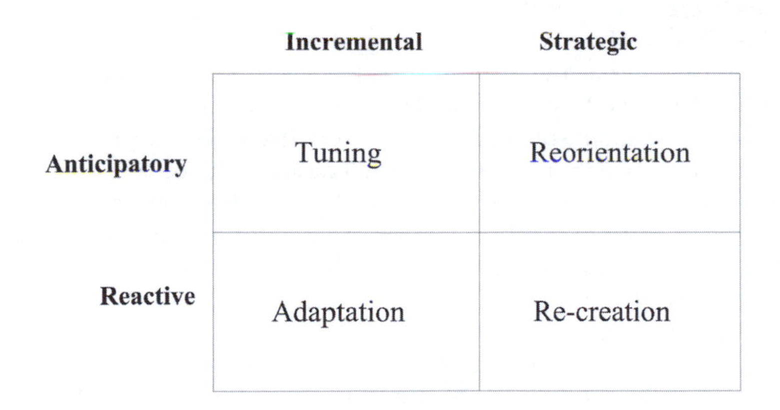

Figure 7.1 Types of organisational change (Jick and Peiperl, 2003).

Theoretical approaches to organisational change

This section provides an overview of four theoretical approaches to change which have contributed to the evolution of organisational change from different perspectives.

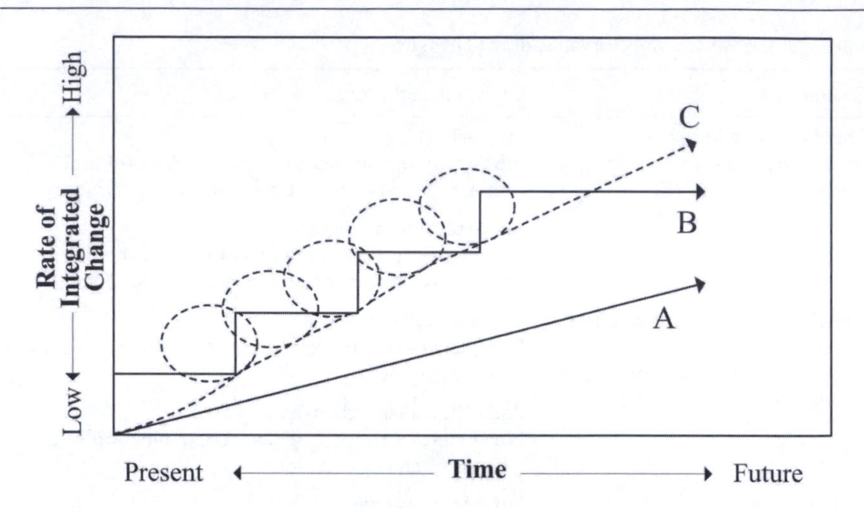

Figure 7.2 Approaches to change (Felkins *et al.*, 1993).

1 *The scientific management approach (Taylorism)*
The basis of the scientific management approach is that work could be divided into sub-units or specialisations. It is believed that there is 'one best way' to perform a task and each sub-unit of a task should be carried out by people capable of carrying out this task in this standardised 'one best way'.
2 *The human relations theory*
The theory dates back to the studies of Elton Mayo and Fritz Roethlisberger at the Hawthorne plant in Chicago during the 1920s and 1930s (Roethlisberger and Dickson, 1939). The Hawthorne studies followed a humanistic approach, drawing attention to group behaviour, relations among group members, and relations between group members and management. Understanding the linkages between the individual, their role among the other members of the group at the workplace and the degree of independence given to the group is considered an initial step to effective performance.
3 *Contingency theory*
Contingency theory is distinguished from most organisational theories proposing 'one best way' to structure an organisation (Donaldson, 2001). According to contingency theory, each organisation has a different way of structuring itself and this structure depends on the circumstances, referred to as contingencies, such as environment, organisational size, technology and organisational strategy (Burns and Stalker, 1961; Chandler, 1962; Child 1973). The theory uses the logic that each organisation has different contingencies, and high performances will occur when the organisational characteristics fit these contingencies (Robey and Sales, 1994). Organisations try to avoid misfits which cause loss of performance; therefore, they adapt themselves

according to the changing contingencies so that effectiveness is maintained. In other words, the will to fit the organisational characteristics to the contingencies results in organisational change.

Burns and Stalker (1961) identify two types of organisations which are effective under different circumstances: mechanistic organisations and organic organisations. The characteristics of the mechanistic and organic organisations are shown in Table 7.4. On the other hand, Lawrence and Lorsch (1967) argue that the design, structure and management of an organisation depend on both the internal and external environments the organisation is based in. The more complex the environment, the more centralised and flexible the management needs to be.

4 *Systems theory*
According to systems theory, in order to understand an organisational survival, adaptation and performance, the dynamics of the environment–organisation relationship should be considered as a system (Wilson and Rosenfeld, 1990). From the change management perspective a system can be defined as 'an assembly of components, which are related in such a way that the behaviour of any individual component will influence the overall status of the system' (Paton and

Table 7.4 Characteristics of mechanistic and organic organisations

Mechanistic organisation	*Organic organisation*
Task differentiation and specialisation	Continuous assessment of task allocation through interaction to utilise knowledge which solves real problems
Hierarchy for coordination of tasks, control and communications	The use of expertise, power relationships and commitment to total task
Control of incoming/outgoing communications from the top and a tendency for information to be provided on a need-to-know basis	Sharing of responsibility
Interaction and emphasis placed on vertical reporting lines	Open and widely used communication patterns which incorporate horizontal and diagonal as well as vertical channels
Loyalty to the organisation and its officers	Commitment to task accomplishment, development and growth of the organisation rather than loyalty to officials
Value placed on internal knowledge and experience in contrast to more general knowledge	Value placed on general skills which are relevant to the organisation

Source: Burns and Stalker (1961).

McCalman 2000: 76). Therefore the systems approaches to change relates to many different dimensions of an organisation.

There have been many models developed for organisational change management following the systems approach. Some of these models differ from each other according to the organisational dimensions they relate to. For example, the organisational development approaches focus more on the soft issues and behavioural aspects whereas the intervention models focus on proposing systematic guidelines in clearly defined steps. These models are discussed in detail in the next section.

Organisational change management models based on the systems approach

Intervention models

Intervention models focus on proposing systematic guidelines in clearly identified steps for implementing organisational change. Although each intervention model involves different steps, three main phases are common to all:

1 definition/description phase;
2 evaluation/design/options phase;
3 implementation phase.

Paton and McCalman (2000) proposed a methodology for the analysis and implementation of organisational change; this is called an Intervention Strategy Model (ISM). The basic stages of ISM are shown in Figure 7.3. Similarly, Senior and Fleming (2006) proposed another approach, the Hard Systems Model of Change (HSMC). The stages of both models are shown in Table 7.5 with respect to the three main phases common to all intervention/hard systems.

Organisational development model

Organisational development (OD) is another model for change which follows the systems approach. However, unlike the intervention models, OD is more related to the soft aspects of change and uses behavioural science technologies, research and theory. Organisational development can be defined as a planned process of change to achieve and improve organisational effectiveness through a systematic application of behavioural and social science methodologies and techniques with the help of a consultant referred to as a change agent (Warner Burke, 1994; Paton and McCalman, 2000; Cummings and Worley, 2005). The change agent is the person who is responsible for the effective implementation of change. The terms 'problem owner', 'facilitator',

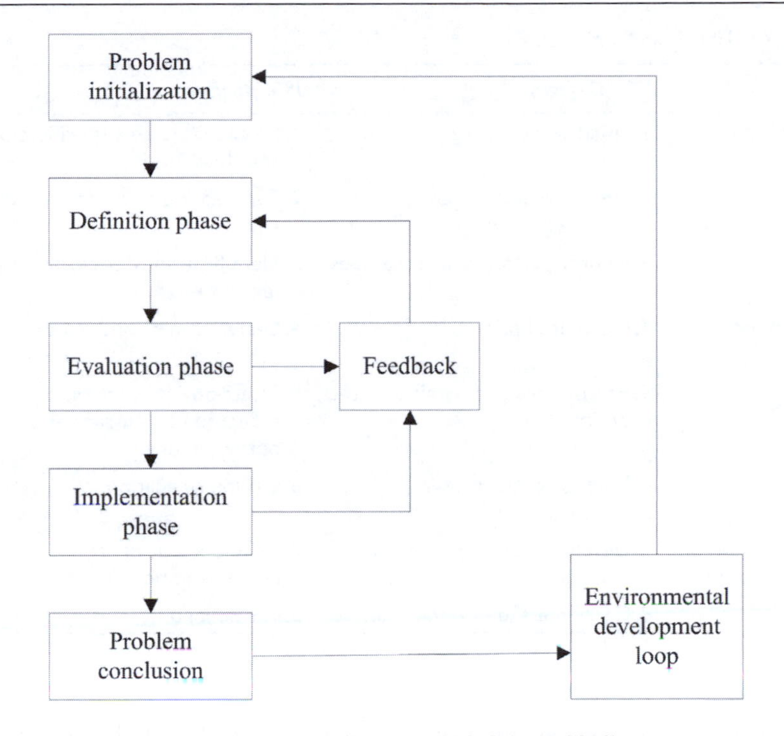

Figure 7.3 The basic phases of the ISM (Paton and McCalman, 2000).

'project manager', 'master of change' and 'change champion' are also used to refer to the change agent (Paton and McCalman, 2000).

The OD approach to change cares about people and believes that people at all levels throughout an organisation are drivers and engines of change (Senior and Fleming, 2006). It considers the organisation as a whole as well as its parts and uses action research as an intervention model. Action research is a collaborative effort between leaders and facilitators of any change and those who have to perform the change (Senior and Fleming, 2006).

Paton and McCalman (2000) define four situations where OD is needed:

1 The current nature of the organisation is leading to a failure to achieve objectives.
2 Change is required to react faster to external alterations.
3 The introduction of factors such as new technology requires change in the organisation itself.
4 The introduction of change allows a new approach to be adopted.

Lewin's early OD model divided change into three phases (Graetz *et al.*, 2006):

Table 7.5 HSMC and ISM stages

Phase	*HSMC stages*	*ISM stages*
Definition	1. Situation summary	1. Problem/systems specification and description
	2. Identify objectives and constraints	2. Formulation of success criteria
	3. Identify performance measures	3. Identification of performance measures
Evaluation	4. Generate options	4. Generation of options or solutions
	5. Edit options and detail selected options	5. Selection of appropriate evaluation techniques and option editing
	6. Evaluate options against measures	6. Option evaluation
Implementation	7. Develop implementation strategies	7. Development of implementation strategies
	8. Carry out the planned changes	8. Consolidation

Sources: Paton and McCalman (2000), Senior and Fleming (2006).

1 unfreezing – raising an awareness of the need for change in the organisation;
2 moving – the stage in which the actual changes are made to move the organisation to a new state through recognition and acceptance;
3 refreezing – integrating and adopting the change through new norms and new ways of working.

This classic three-stage change model has been superseded by more detailed adoption models which include additional feedback and evaluation stages. According to Warner Burke (1994) an organisation will experience seven phases during a typical OD change process:

1 entry phase;
2 formalising the contact;
3 information gathering and analysis;
4 feedback;
5 planning the change process;
6 implementing the changes; and
7 assessment.

The OD model developed by Senior and Fleming (2006), shown in Figure 7.4, is another example that includes feedback and correction loops between stages. The feedback and correction loops led to the development of the learning organisation concept, which is now considered one of the core principles of OD. There are two types of organisational learning: single loop learning and double loop learning. These are illustrated in Figure 7.5. Single loop or instrumental learning is an adaptive learning through which an entity learns to do better than it is currently doing (Senge, 1992; Paton and McCalman, 2000). Incremental change and adaptation by means of Total Quality Management (TQM) are examples of single loop organisational learning approaches.

Double loop learning or generative learning aims to challenge long-held assumptions and to create new ways of looking at the world; therefore, it not only alters the decisions made for the organisation but also feeds back to the mental models of the real world (Argyris, 1985; Senge, 1992; Paton and McCalman, 2000; Sterman, 2000).

Senge's (1992) work on learning organisations is considered to have led to the development of the five disciplines aimed at enhancing an organisation's creative capability: personal mastery, mental models, building a shared vision, team learning and systems thinking.

Tools used to introduce organisational change in construction organisations

The organisational change management concept has strong links with human resource management, risk management, organisational learning,

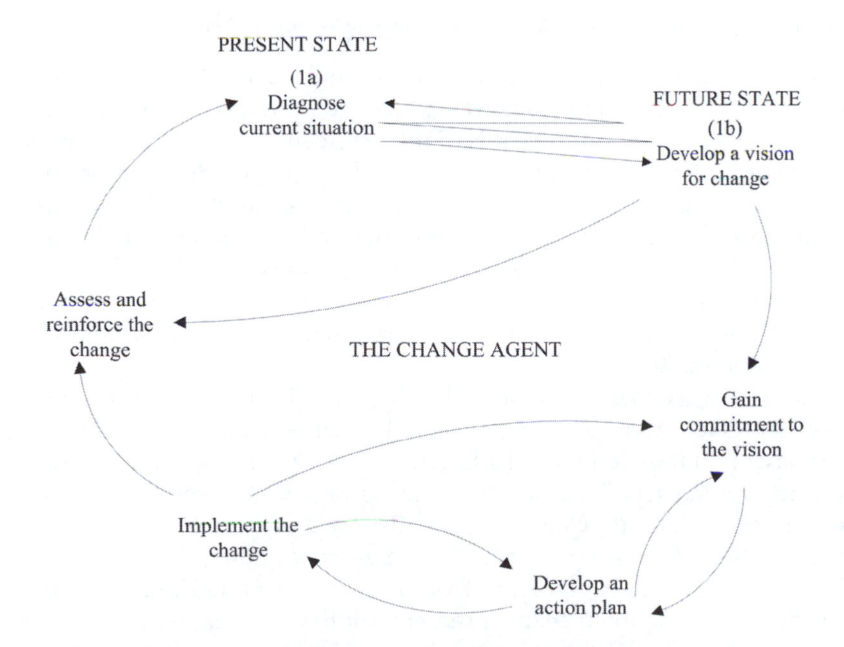

Figure 7.4 The OD model for change (Senior and Fleming, 2006).

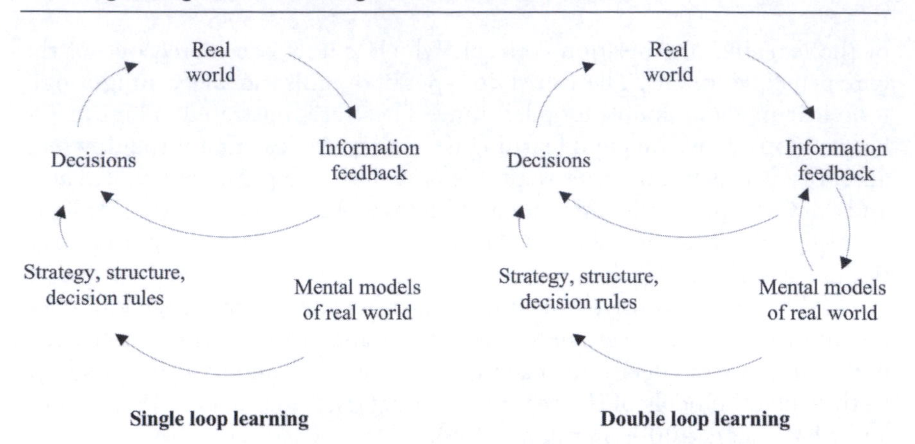

Figure 7.5 Single and double loop learning (Sterman, 2000).

strategic management, information technology management and quality management concepts, and overlaps with organisational development and organisational dynamics. Some organisational changes are known by their specific names according to their focus and level. Business process re-engineering (BPR) and TQM are examples of these changes. Although these changes are introduced at different levels, they require a well-organised and -planned change management implementation in the organisation.

Introducing change through business process re-engineering (BPR)

As stated in the reports by Latham (1994) and Egan (1998), the pressure on the construction sector to increase its productivity and improve quality is growing. Construction organisations are focused on the outcome and success of individual projects, with relatively little consideration of how to achieve the same and preferably better results repeatedly and consistently. To increase the quality of the end product and productivity, they should focus on the processes followed and the elements and the sub-elements constituting the processes. The objectives of BPR are quick and substantial gains in organisational performance by redesigning the core business processes (Attaran, 2000).

There is a lack of common definitions for BPR, and other types of concepts related to it. This has reflected negatively in organisational perceptions of BPR practices (Al-Mashari *et al.*, 2001). It is also criticised by many researchers because of its limitations in dealing with change at organisational level (O'Conner, 1994; Ruessman *et al.*, 1994; Vakola and Rezgui, 2000; Al-Mashari *et al.*, 2001; Cao *et al.*, 2001). It is not possible to isolate process re-engineering from the structural, cultural or political aspects of organisational change; therefore it has been suggested that either the use of BPR be limited to those process-driven situations, or a holistic view be introduced to help deal with a wider spectrum of change (Cao *et al.*, 2001).

There are several barriers to successful re-engineering implementation; these include poor top management support and involvement, lack of flexibility, lack of effective organisational communication, lack of proper training, failure to cope with people resistance, failure to assign adequate human resources to re-engineering efforts, misunderstanding and misapplication of the concept, and failure to test the process (Attaran, 2000). Although there have been some recent improvements in dealing with human and process issues, BPR use still fails to provide the organisational change management expectations as its scope does not go beyond process level.

Introducing change through maturity models

The maturity concept originated in the quality principles defined by Philip Crosby (1994) as five evolutionary stages in adopting quality practices. These were subsequently modified by the software industry (Humphrey 1987, 1988) and resulted in the capability maturity model (CMM). CMM describes five levels of maturity for software process improvement. The maturity of the organisation increases through each level and each maturity level provides a step in the quest for continuous process improvement (Figure 7.6).

Each level comprises a set of goals that, when satisfied, completes an important component of the process, resulting in an increase in the process capability of the organisation (Paulk, 1993). The conditions of the five levels of CMM are briefly explained below (Humphrey 1987, 1988; Paulk, 1993):

1. *Level 1: initial/ad hoc*
   Project visibility and predictability are poor; there is an unstable environment for developing and maintaining products; delays in the project time and cost overruns occur frequently; successes depend on the individual efforts rather than those of the team or organisation.
2. *Level 2: repeatable*
   The organisations tend to meet their schedule commitments but cost is not as controllable as the schedule; the organisation has some policies for managing projects and has established a structure to implement these policies.
3. *Level 3: defined*
   The organisations in this level have standard processes defined and allocated resources for developing, sustaining and improving these processes; the organisations are supposed to have stable and repeatable cost, schedule and quality systems; some organisation-wide training programmes are implemented.

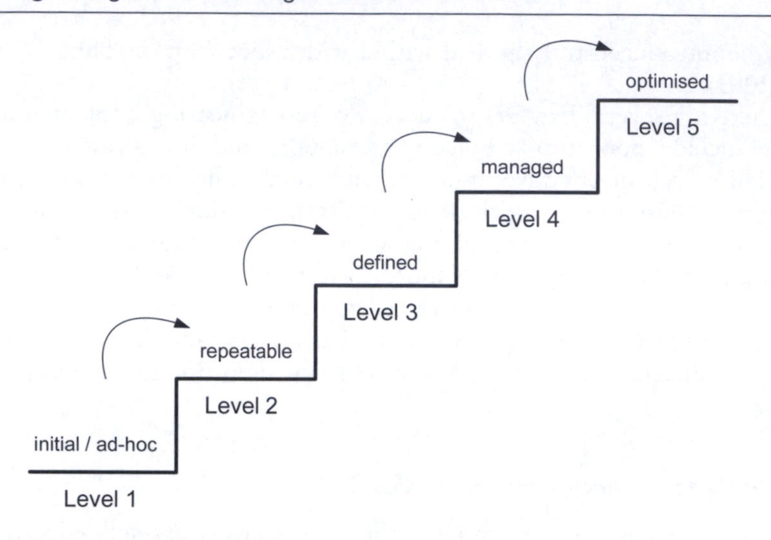

Figure 7.6 Maturity levels in CMM.

4 *Level 4: managed*
The organisations generally meet or exceed the defined quality goals by operating within predictable quantitative performance levels; process measurement systems are also established at this level.

5 *Level 5: optimised/optimising*
This is the optimum maturity level for organisations. The improvement goals in this stage are established; the organisations' objectives become aligned with 'continuous process improvement'; regular defect prevention methods are carried out and weaknesses are determined and eliminated in a continuous process.

Introducing change through TQM

Total quality (TQ) is defined as a way of meeting customer requirements whereby the customers are both internal and external to the organisation. It comprises a change invoked through four key components (Rye, 1996):

1 *Systems*: The systems refer to the online and offline quality procedures whereby the quality of change is supported with quality control, quality assurance and foolproofing.
2 *Processes*: TQ regards every activity of an organisation as a part of a process. Therefore, it encourages the constant review of processes for continuous improvement, waste elimination and process chain re-engineering.
3 *People*: TQ companies value their employees as individuals and also for their contribution to the growth of the company.

4 *Management*: The management concept in TQM is related to the vision and mission, critical success factors, organisation for quality, championing and empowerment.

When organisations implement TQM, they become engaged in, inter alia, continuously improving operations, meeting customer requirements, reducing rework, thinking long range, increasing employee involvement, redesigning processes, conducting competitive benchmarking, measuring results continuously and fostering closer relationships with suppliers (Singh and Smith, 2004).

Irani and colleagues (2004) discuss the concept of corporate culture, and place this social construct within the arena of TQM to conclude that the core concept of TQM together with the customer focus, linked with a continuous improvement plan supported by innovation, can build a robust culture which can positively improve an organisation's competitiveness and performance.

The main aim in TQM is to improve the organisation without making major changes; therefore, TQM has a high potential in achieving radical results but fails in reaching solutions beyond information management. Thus, TQM is not considered a very efficient organisational change management tool.

Leading the change

Leadership is a very important dimension in organisational change management and therefore those in charge of change should be carefully chosen. The person leading the change can be referred to as a change agent, problem owner, problem facilitator, master of change or change champion (Warner Burke, 1994; Paton and McCalman, 2000; Cummings and Worley, 2005).

Most changes are introduced through the analyse–think–change approach, which is regarded as the best scientific approach, whereby data related to a specific change requirement are collected and analysed, and then an appropriate solution is selected and implemented. However, some argue that scientific approaches may not always be the best way to manage change involving people. Kotter and Cohen (2002) list three limitations starting with the analysis stage:

1 In most cases detailed analysis reports are not necessary to understand that something is not working.
2 Analytical tools work best when parameters are known, assumptions are minimal and the future is not fuzzy.
3 Good analysis changes the thoughts of people but does not motivate them in a big way.

Kotter and Cohen (2002) suggest that in order to get people involved in the changes, or to get them to follow the changes introduced, the see–feel–change approach works better than the analyse–think–change method. The see–feel–change approach is targeted at helping people see by visualising the problems or solutions through the use of eye-catching, compelling and dramatic situations (Kotter and Cohen, 2002). These visualisations provide useful ideas which will stimulate emotions to change behaviour or reinforce changed behaviour (Kotter and Cohen, 2002).

Eight stages are proposed by Kotter and Cohen (2002) in order to achieve a large-scale change using this see–feel–change approach. Table 7.6 explains these eight stages and the new behaviours created by the actions at each stage.

Another framework for managing organisational changes is proposed by Jick and Peiperl (2003). This framework consists of 10 stages named 'The ten commandments of implementing successful organizational change'. These 10 stages are:

1 Analyse the organisation and its need for change.
2 Create a shared vision and common direction.
3 Separate from the past.
4 Create a sense of urgency.
5 Support a strong leader role.
6 Line up political sponsorship.
7 Craft an implementation plan.

Table 7.6 Kotter's eight stages to achieve a large-scale change

Step	*Action*	*New behaviour*
1	Increase urgency	People start telling each other, 'Let's go, we need to change things!'
2	Build the guiding team	A group powerful enough to guide a big change is formed and they start to work together well
3	Get the vision right	The guiding team develops the right vision and strategy for the change effort
4	Communicate for buy-in	People begin to buy in to the change and this shows in their behaviour
5	Empower action	More people feel able to act, and do act, on the vision
6	Create short-term wins	Momentum builds as people try to fulfil the vision, while fewer and fewer resist change
7	Do not let up	People make wave after wave of changes until the vision is fulfilled
8	Make change stick	New and winning behaviour continues despite the pull of tradition, turnover of change leaders etc.

8 Develop enabling structures.
9 Communicate, involve people and be honest.
10 Reinforce and institutionalise change.

The 'ten commandments approach' focuses on defining the need for change, creating a vision and the role of a strong leader, whereas Kotter stresses the importance of a team in charge of the change implementation rather than the one person assuming the leadership role.

Resistance to change

It is difficult for organisations to implement change successfully unless it is accepted by their employees. Resistance from employees can be inevitable. In order to manage this resistance, leaders should appreciate the different conditions and priorities of their employees, understand the sources of resistance and try to identify a resolution. Based on this idea the following sections discuss how individuals react to change, the causes and different forms of resistance and how it can be overcome.

Reactions to change

There have been many attempts to model people's reactions to change reactions and the way they cope with it. Carnall (1990) defines five stages of change implementation behaviour: denial, defence, discarding, adaptation and internalisation. The way people's performance and self-esteem vary with time during these five stages is shown in Figure 7.7.

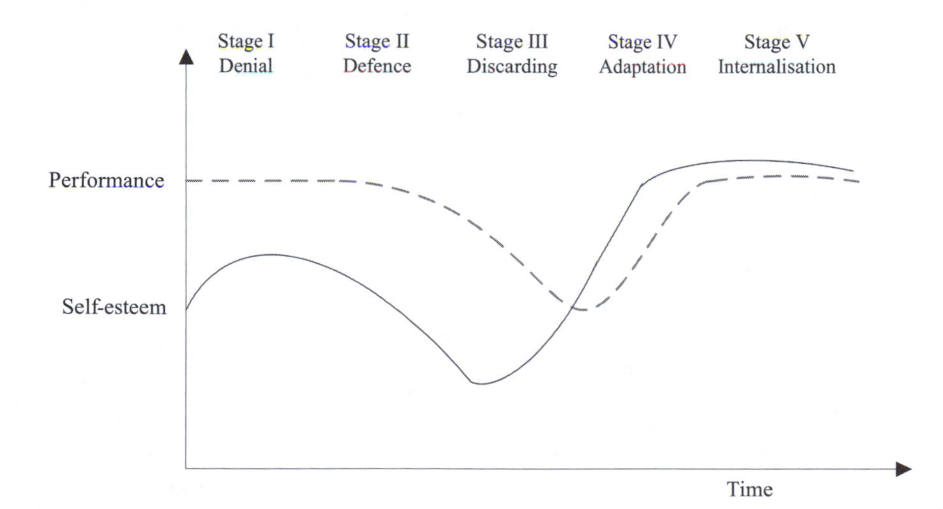

Figure 7.7 The coping cycle (Carnall, 1990).

Conner (1993) suggests that people's response to change varies according to their initial perception, and proposes two different response–time relationships: negative and positive. In the negative version shown in Figure 7.8, the initial reaction is a shock with an immobilisation stage due to the removal of the stable environment. This stage is followed by a denial stage, when the change-related information is often rejected or ignored in the hope that change will fade away. Next follows the anger stage. After the realisation that it is no longer possible to avoid reality, people begin negotiating in order to avoid the negative impact of change. This bargaining process is followed by a depression stage after the full impact of the consequences is finally acknowledged. People get over this depression stage by regaining a sense of control through acknowledging the new limitations and exploring ways to redefine goals. This stage is called the testing stage and is followed by acceptance when people respond to the change realistically.

If the initial reaction to change is positive, the pessimistic states vary with time, as shown in Figure 7.9. When the change is first introduced, it is welcomed by an uninformed modest optimism and low pessimism. The initial realisation of some of the real impacts of the change then leads to an informed pessimism phase, and if this pessimism is high individuals may decide to withdraw, resulting in 'checking out'. If, however, the initial difficulties are addressed then a hopeful realism stage is reached. As more and more problems are resolved, individuals progress to an informed optimism phase. The pessimism level continues to go down throughout the different phases until it reaches a negligible minimum on completion.

Another model of the negative response to change was proposed by Jick and Peiperl (2003), who propose two frameworks to explain individuals'

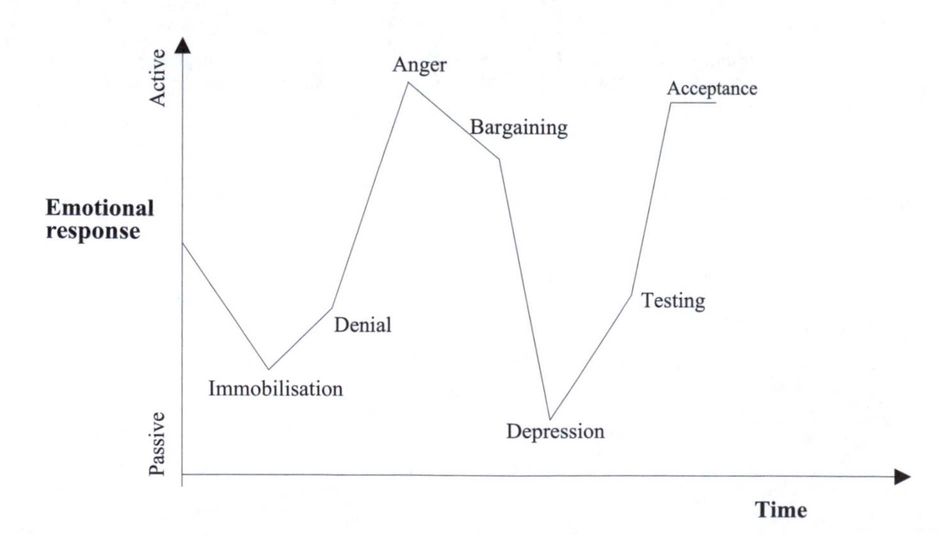

Figure 7.8 Negative response to change (Conner, 1993).

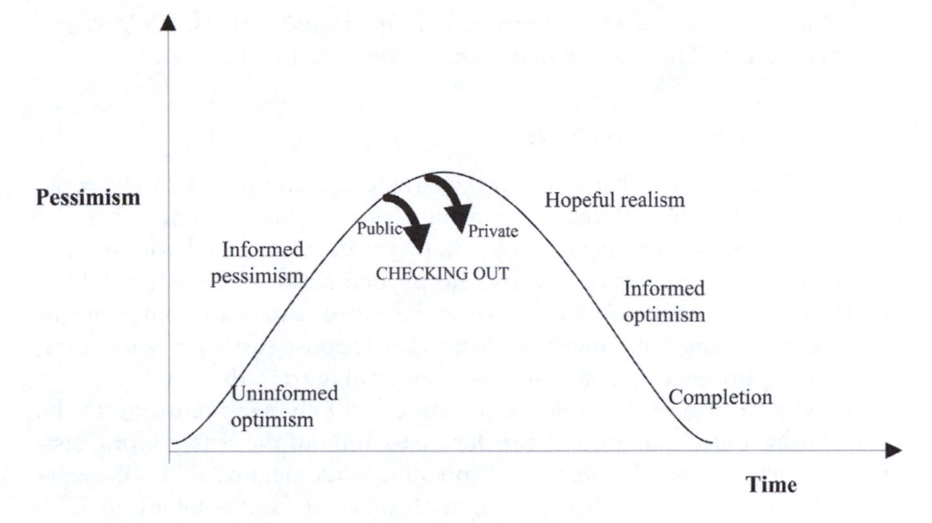

Figure 7.9 Positive response to change (Conner, 1993).

reactions to change. This suggests that individuals pass through four stages during a change implementation: shock, defensive retreat, acknowledgement, adaptation and change. In the shock phase, the individuals feel threatened and become immobilised in an attempt to not take risks. This phase is followed by anger and insistence on holding on to the current conditions. Eventually, the change is acknowledged while the mourning for the lost accustomed ways still continues; then the pros and cons of the change are explored and the risk-taking potential increases. Confidence is built and continues to increase after each risk is overcome. The final phase is the adaptation to change phase, when the individuals become comfortable with the new conditions.

A further model by Jick and Peiperl (2003) captures how people respond and cope with change through three transition stages:

1 *ending phase* – letting go of the previous situation;
2 *neutral zone* – completing the end and building energy for beginning;
3 *new beginnings* – new possibilities or alignment with a vision.

Understanding the stages individuals go through during a change is an important step towards dealing with resistance. Another one is to understand the reasons behind the employee resistance.

Reasons for employee resistance

Resistance to change is a result of perceived differences in ideas, motives, plans or priorities (O'Connor, 1993). Resistance to change is found to be related to five areas: need for change, risk, goals and targets, leaders, and threats to status. The reasons in each area are shown in Table 7.7.

Different behaviours during resistance

Since each individual has different characteristics, the way they resist change will also be different. Behaviours can be based on whether the employees express their feelings openly (overt or covert) and whether the individuals are conscious or unconscious of their actions. O'Connor (1993) defines four different behavioural types based on different combinations of overt/covert and conscious/unconscious reactions: survivors, saboteurs, zombies and protesters. These are explained in Figure 7.10.

The way to deal with employee resistance will change according to the type of behaviour. Starting an open debate to find out the reasons for resistance and overcoming the obstacles is possible when dealing with saboteurs and protesters whereas, when dealing with survivors and zombies, managers should find ways to help them understand what they are doing and why. Often, the resistors are also as aware of a need for change as others but may have their own ideas on how it should be managed. When implementing organisational change, resistance can be managed by following five key principles (Conner, 1993):

Table 7.7 Reasons for employee resistance

Area	*Reasons for resistance*
Need for change	Lack of belief or understanding of the need for change Lack of perceived benefits Different descriptions of the need for change Misunderstanding of the change
Risk	Fear of the unknown Uncertainty regarding the change outcomes Fear of failure
Goals/targets	No agreement about goals for change Lack of belief that the goal is attainable Belief that the proposed change is not aligned with company's core values Belief that change is not in the best interests of the organisation Lack of, or a different sense of, the context or environment
Leaders	Lack of trust in the people introducing or managing the change Lack of belief that leadership is serious about making changes Lack of belief that the leadership is capable of making change happen
Fears of status quo	Perception that the change is unfairly selective Threats to status Internal politics (such as elitism and interdepartmental rivalry) Lack of information/knowledge/skill

Sources: O'Connor (1993), Quirke (1996), Ford *et al.* (2002), Proctor and Doukakis (2003).

1 Understand the basic mechanisms of human resistance.
2 View resistance as a natural and inevitable reaction to the disruption of expectations.
3 Interpret resistance as a deficiency of either ability or willingness.
4 Encourage and participate in overt expressions of resistance.
5 Understand that resistance to positive change is just as common as resistance to negatively perceived change and that both reactions follow their own sequence of events, which can be anticipated and managed.

Staff leading the change should always remember that the resistance of employees is directed at the change, not at the person implementing the change, and should therefore mange it objectively. Reacting to change forcefully will create two poles so that nobody will give way to the other. Communication of change also plays a very important role in managing employees' resistance.

Communicating change

Plans for change are often made by one company group and then implemented by a different one, traditionally lower in the management hierarchy,

COVERT

UNCONSCIOUS	CONSCIOUS
SURVIVOR ▪ do not realise they undermine the change ▪ do not know they fail to meet targets ▪ do not understand the implications of their behaviour ▪ do the job in the way they know how to do ▪ are difficult to detect due to the higher profile projects masking them ▪ are as surprised and disappointed as anyone in the management, when their lack of adaptation to change is discovered	**SABOTEUR** ▪ undermine change while pretending to support it ▪ believe not doing anything other than verbal support will make the change initiative go away / disappear ▪ intend to sabotage the plan for their own benefit ▪ are very common in highly competitive environments
ZOMBIE ▪ are extreme cases of survivor ▪ verbally agree to do whatever is asked of them, they have neither the will nor the ability to create change ▪ gradually and openly they revert back to previous behaviour ▪ simply avoid the change until they are reminded again to alter their behaviour	**PROTESTER** ▪ believe that their refusal to change is a positive contribution to the company ▪ never give up pointing out the failings of change ▪ are the easiest and the most interesting kind of resisters to manage since it is possible to discuss their position clearly and rationally

OVERT

Figure 7.10 Categories of behaviour: a matrix (adapted from O'Connor, 1993).

with minimum communication between the two. The separation of these two activities generally results in a communication breakdown leading to misunderstandings and failed aspects of implementation (O'Connor, 1993). Therefore, communication of change and its timing play a vital role in the success of organisational change management since it encourages the employees to adjust.

The effective early communication of change to employees should be based on four principles (Balogun and Hope Hailey, 2004):

1 Employees prefer to find out about change from the management rather than through informal channels.
2 Early communications allow employees time to understand and adjust.
3 Employees prefer honest and even incomplete announcements to cover-ups.
4 Employees learn about changes despite policies of silence.

Paton and McCalman (2000) proposed five-step guidelines to be followed for communicating change:

1 customise the message;
2 set the appropriate tone;
3 build in feedback;
4 set the example; and
5 ensure penetration.

The nature of information, appropriateness of communication media and the likely consequences of these should be considered before communicating with employees (Weiss, 2001). Three different methods to communicate changes are usually recommended:

1 a general announcement on a notice-board or an open staff meeting;
2 an individual memo to each person affected by the change;
3 person-to-person communication for important changes.

The medium for the first two methods is most likely to be a form of electronic communication. Whichever medium or method is chosen, the selected method should suit the company/group culture or work habits and enable the capture of feedback. For example, newsletters, brochures and videos are very common tools used to inform the employees of changes. However, they cannot be considered effective, as change communication is not just a one-way information flow; it is mutual information exchange that should provide opportunities for feedback and debate (Quirke, 1996).

Chapter 8

Change management framework

Bilge Erdogan, Dino Bouchlaghem and Yasemin Nielsen

Introduction

As explained in the previous chapter, the success of a collaboration environment not only depends on what is introduced to the organisation but is also highly influenced by how it is introduced. When a new system is introduced in an organisation, it results in the adoption of new technologies that will inevitably affect some work processes. Therefore, the organisation has to adapt these processes to enable the effective implementation and adoption of the new technological environment. Thus, it is important that the introduction of new collaboration environments is managed through an organisational change management approach.

This chapter presents a structured methodology for collaboration implementation, together with a detailed organisational change management approach that addresses factors affecting the success of collaboration environments. The methodology is presented in the form of a framework, ICEMOCHA (Implementation of Collaboration Environments and Management of Organisational CHAnge), that aims to support collaboration in construction projects with a focus on organisational change management.

The ICEMOCHA framework can be described as a specific implementation of the generic Planning and Implementation of Effective Collaboration in Construction (PIECC) framework (see Chapter 3). Although both of them work towards the same objective, they are very different in terms of motives, perspectives, target levels and representation methods. PIECC follows a strategic management perspective whereas ICEMOCHA approaches the problem from a change management perspective. Therefore, while PIECC provides guidance in planning a collaboration strategy, ICEMOCHA focuses on implementing collaboration environments and managing the related organisational changes. Unlike the PIECC project, which is developed for use at strategic level, ICEMOCHA applies to tactical and operational levels.

The chapter first gives an overview of the background and rationale for the framework, its aims and objectives, and the development approach used. This is followed by a detailed description of the framework in which each process is explained.

Overview of ICEMOCHA and background for its development

The development of the ICEMOCHA methodology was based on the PIECC framework presented in Chapter 3. The methodology can be described as a focused implementation of PIECC. Whereas PIECC adopted a strategic management perspective to provide guidance for the planning of collaboration, ICEMOCHA deals with the operational implementation of collaboration and the management of the associated organisational changes required. The following are the background considerations that led to the development of the methodology:

- Both previous research and the case studies presented in Chapter 9 indicate that the reasons for the failure of the collaboration environments are not technological but mostly related to organisational and human issues.
- The key factors to be considered when implementing collaboration environments have been identified as user requirements capture, overcoming user resistance to change, user involvement, effective planning/project management, strategic information technology (IT) implementation, buy-in from all parties and trust. The case study results also supported these findings together with the PIECC project findings presented in Chapter 3.
- The success of collaboration environments does not only depend on what is introduced to the organisation but is also dependent on how it is introduced. ICEMOCHA provides a methodology to manage how the collaboration environments are introduced within organisations.
- The construction industry has invested little effort in organisational change management when implementing information technologies or collaboration environments.
- The grouping of organisations participating in a construction project and the dynamics between them can be considered to form a system, and failure of one company in the adoption of collaboration environments would affect this system as a whole. Therefore systems theory can be adopted for the strategic management of collaboration environment implementation and organisational change.
- When a new collaboration environment is introduced in an organisation, it results in change within two contingencies: a new working approach and a new technology. The organisation has to adapt to these new contingencies. Therefore, the introduction of new collaboration environments should be managed through an organisational change

management approach in order to adapt the organisation to the changing contingencies.
- The case studies indicated that the factors that affect the success of collaboration environments are present at both project and organisational levels.

Based on these underlying principles, the specific aims and objectives of ICEMOCHA were defined and are presented in the next section.

Aims and objectives of ICEMOCHA

ICEMOCHA aims to support the effective implementation of collaboration environments in construction projects and includes a methodology that guides the organisational change management required. The specific objectives of ICEMOCHA are:

- to manage the changes brought into construction organisations by the introduction of a new collaboration environment;
- to increase attention on the people and organisational issues in the planning and implementation of collaboration technologies;
- to prevent/manage resistance to change and to cope with other barriers to the implementation of collaboration environments;
- to improve collaboration across construction projects by improving the efficiency of the collaboration environment and collaboration tools, considering different dimensions such as strategy, technology, organisational processes and people.

The framework is for use by construction organisations working on collaborative projects using modern IT tools to create a collaboration environment. It is intended for application at senior level but can also be used by middle-level management. Users of ICEMOCHA can include business managers, project managers, IT managers, research and development department managers and employees, collaboration champions, team leaders and a sample of end users.

Conceptual framework

The ICEMOCHA framework is based on the principle that the success of collaboration in projects requires all organisations involved to participate in the collaboration environment and follow the same procedures. Each organisation needs to manage the changes brought about by the new collaboration environment in order to adjust their existing processes and ensure that their employees use the new systems efficiently. The amount of the change required will depend on how familiar the organisation is with the new collaboration technologies, tools or methodologies, and their level of compatibility with the new organisational processes and working culture.

There are two levels that should be considered during the implementation of a collaboration environment in a construction project: project organisation level and organisational level. ICEMOCHA applies to both these levels, as shown in Figure 8.1.

'Project organisation' is a term used in ICEMOCHA to refer to the virtual temporary organisation formed when all organisations collaborating come together to make decisions regarding the overall project. Project organisation-level decisions are usually made in the presence of representatives from each organisation; the agreed collaboration solution and related decisions are therefore binding for all organisations. These procedures should then be adopted by each organisation, ensuring that these common decisions are known to and followed by all employees. However, in order to achieve an efficient adoption, an organisational change management approach is required. The organisational-level processes of ICEMOCHA provide a methodology which enables each organisation to come up with an organisation-specific change management approach. These processes should be carried out by each organisation individually since the changes required will be specific to the different organisational cultures and varying organisational procedures and processes.

ICEMOCHA is a combination of two interlinked process models: the implementation of collaboration environments (ICE) model for use at project organisation level, and the management of organisational changes (MOCHA) for use at organisational level. An overview of the ICEMOCHA framework is shown in Figure 8.2. Both models follow a scientific problem-solving approach that involves five stages. These steps are used for collaboration management at the project organisation level and for change management at the organisational level. The first stage is referred to as the initiation stage, where the need for collaboration and for organisational

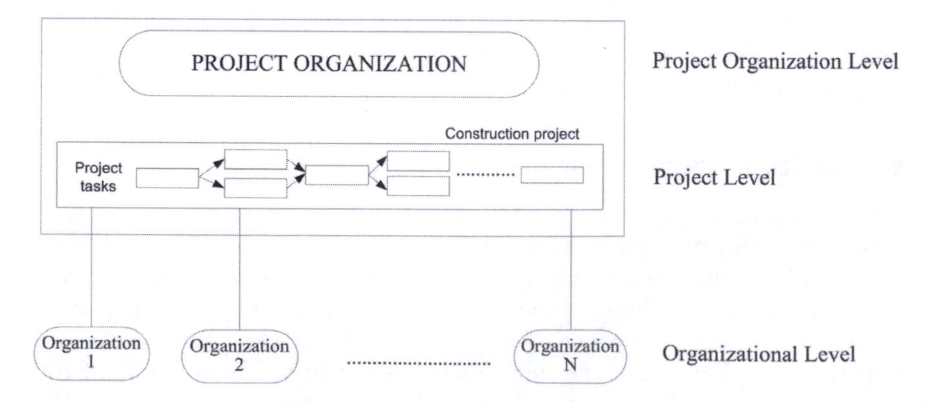

Figure 8.1 The ICEMOCHA levels.

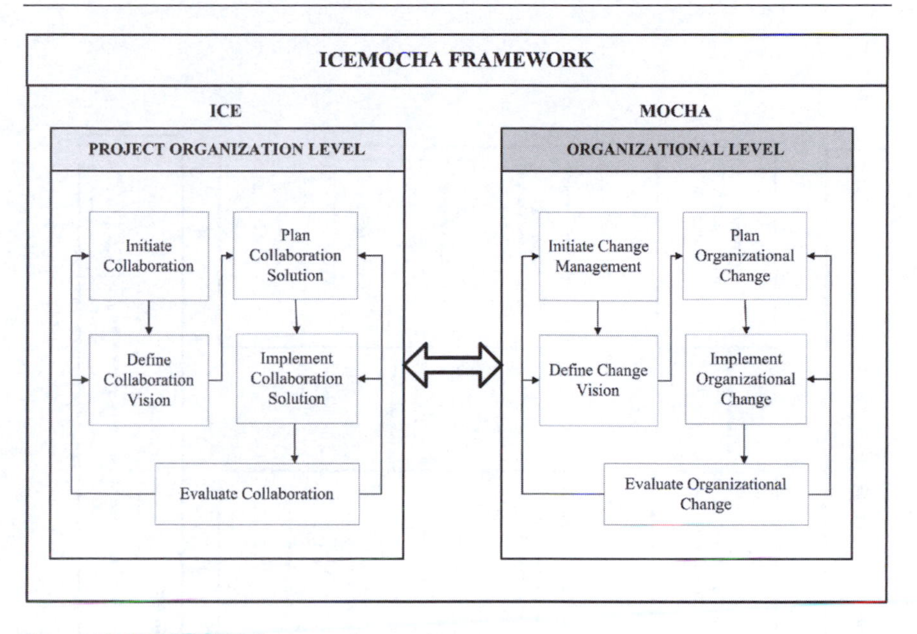

Figure 8.2 ICEMOCHA framework overview.

change is defined; while the second stage focuses on defining the vision. In the ICE model a shared collaboration vision is developed, whereas in the MOCHA model a change vision is defined. In the third stage, a collaboration solution and an organisation change management plan are required in ICE and MOCHA respectively. The fourth stage focuses on implementing the solutions, while the final stage focuses on evaluating the performance of the implementation. A detailed representation of the overall ICEMOCHA framework is shown in Figure 8.3.

Actors in the framework

The key responsibilities for change management are carried out by the collaboration champion appointed as part of the PIECC framework in Chapter 3. The collaboration management team is also put together in a similar way to that described in the PIECC framework. The various actors are described in detail below.

Project director

The project director is the overall leader of the construction project and is responsible for the high-level strategic decisions that affect the whole project. He/she communicates with all parties through the collaboration champions to oversee the coordination of all project activities.

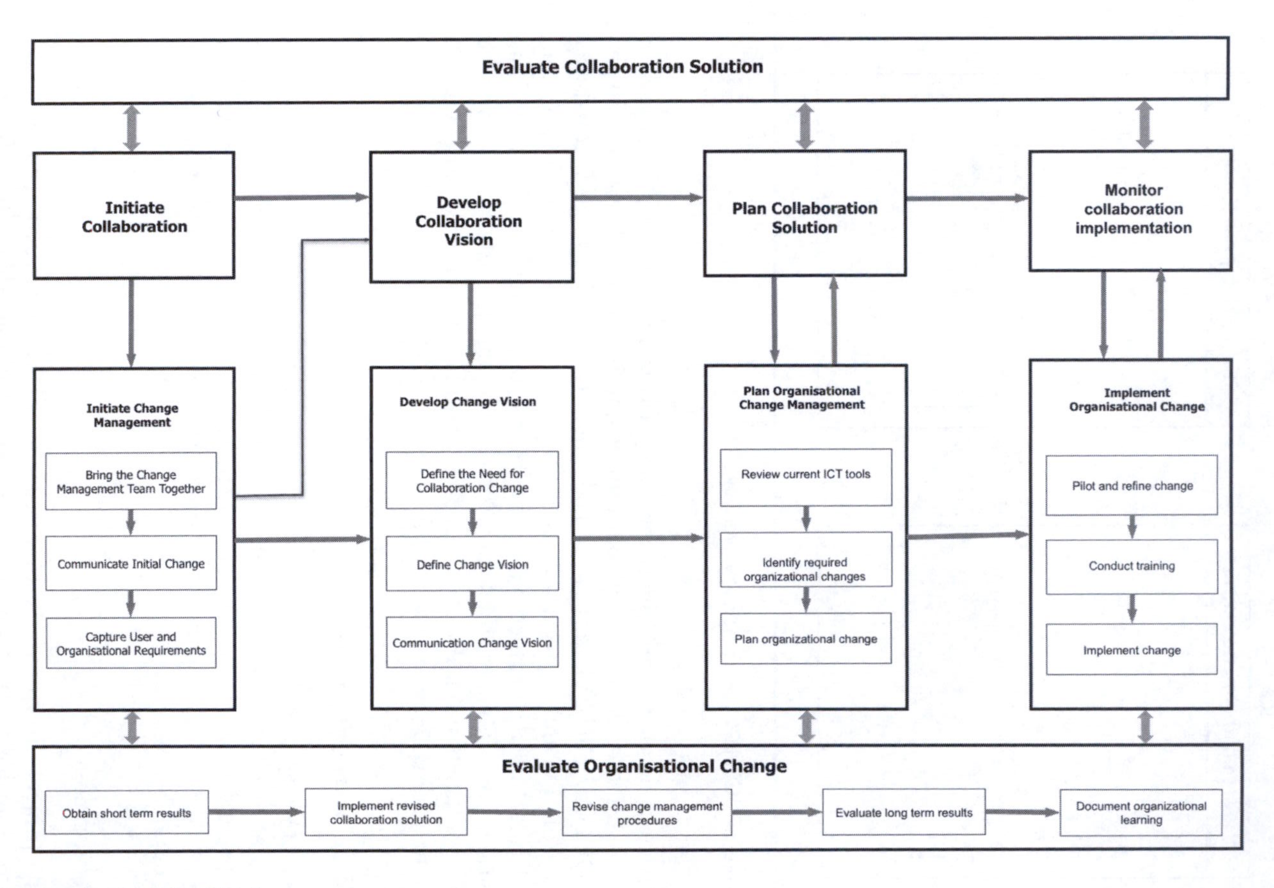

Figure 8.3 Detailed ICEMOCHA framework.

Collaboration champion

Each collaborating organisation appoints a suitable champion as its representative in the collaborative venture. These champions are responsible for all collaboration-related tasks within their organisation. They manage the implementation of collaboration environments and ensure that their organisation participates in the project according to the collaboration standards and procedures defined by the project team. They are also responsible for managing the changes required by the collaboration environment and should therefore aim to achieve a smooth adoption of the collaboration systems working with the collaboration change management team members. They are also responsible for the coordination of all collaboration activities throughout the project, leading the collaboration change management team and working together with the IT/systems manager.

Collaboration management team (CMT)

The CMT is essentially a decision-making board that manages the collaboration at the project organisation level. The team is responsible for decisions related to the collaboration implementation that apply to all participating organisations. The smallest CMT would consist of the collaboration champions from all parties involved. Since the collaboration-related decisions will also affect technologies and processes at different levels and within different departments, the CMT should ideally include representatives from each organisation with good knowledge of all related areas. These representatives might be IT personnel and/or others familiar with relevant processes affected by the collaboration venture. The CMT might also include some of the collaboration change management team members (see below).

Collaboration change management team (CCMT)

The CCMT consists of a sample of users from each organisation appointed by the collaboration champion and/or senior management to enable the link between the users and the collaboration champion. The team might include representatives from operational, technology and senior management levels.

The CCMT is one of the key elements of the organisational change management framework and is a result of a user involvement approach. The team plays an active role in capturing user and organisational requirements, and defining and communicating changes within the organisation. The team also participates in all collaboration implementation and evaluation processes. Having representatives from those affected by the change will make its adoption easier as it will no longer be seen as something imposed. Being given the opportunity to discuss the change directly or indirectly through the CCMT will give the users the feeling of ownership.

Senior management

'Senior management' refers to a team of high-level managers responsible for the day-to-day activities of an organisation. Senior management appoints a collaboration champion to represent their organisation in the project. They may also decide to select the CCMT members or alternatively delegate it to the collaboration champions. The CCMT will also include members from senior management; this will ensure top-level commitment during the implementation and hence help to decrease user resistance and increase the level of uptake of the collaboration environment.

Systems manager

A systems manager is a senior manager who is in charge of all IT activities and facilities within an organisation. They are also referred to as 'Information Systems Manager' or 'IT Manager' in some organisations. If there is a specific IT department, the systems manager will be the head, leading technical staff such as programmers, database administrators and systems analysts.

Unlike the traditional approach, whereby the systems managers are fully responsible for the introduction of collaboration environments and are therefore involved in all stages, in ICEMOCHA, the systems managers' role is limited to the implementation and evaluation of collaboration environments. However, the CCMT may include the systems manager or some IT staff to address the technical issues during the requirements capture, collaboration vision and solution-planning stages.

Description of the ICE process

This process is mostly based on the PIECC framework described in Chapter 3 and adapted to suit the project-level context.

Initiate collaboration

The project contract should include clauses that define the need for collaborative working with detailed procedures for collaboration. If there are no binding clauses within the contract, then the project director should establish whether the use of collaboration tools is critical for the success of the project. Feasibility studies for the use of collaboration environments should therefore be carried out. When the need for IT-enabled collaboration is established, the project director should develop the preliminary collaboration specifications taking into account the project and technical requirements. These specifications are then communicated to other organisations participating in the project so that they can identify any changes needed to suit their own organisational requirements for collaboration.

The project director and collaboration champions should then set up the CMT. They can request the appointment of specific members from each organisation or inform them of the profile of staff needed and let them select their own representatives. It is important that the CMT and CCMT have overlapping membership.

Develop collaboration vision

Having an adopted shared vision that takes into account the characteristics of all organisations involved at project organisation level will make it easier for the parties to agree on subsequent actions regarding the collaboration. Collaboration risks should then be assessed by the CMT using the preliminary collaboration specifications and the shared collaboration vision (for risk assessment techniques see Chapter 3). During this process, the collaboration risks are identified and, for each risk, the probability of occurrence, severity of impact and mitigation actions should be outlined.

Plan collaboration solution

This task aims to define a collaboration solution to be agreed and accepted by each organisation for use in the project. The CMT plays a leading role in this process but it is important to also include other representatives familiar with the processes and working methods that are likely to be affected by the collaboration. They can then ensure, as much as possible, that any new systems and procedures considered are in line with their own organisations' working methods. Although it is not always possible to accommodate the requirements of all parties, it is important to create an environment where the needs of all are considered and compromises are made to arrive at a best-fit solution. The CMT members are as such well placed to promote the collaboration solution agreed within their organisations and hence ensure the success of the collaboration venture. The collaboration solution usually covers four main areas as outlined in the PIECC framework: definition of roles and responsibilities, project lifecycle data/work flow, technology specifications, and collaboration standards and procedures.

Monitor collaboration implementation

This is based on the PIECC framework and aims to monitor the implementation of the collaboration environments within the collaborating organisations through various processes including the recruitment of participants, the purchase of collaboration tools and technologies, the provision of training, the piloting of collaborations solutions and the final implementations. The most important task in this process is the recruitment of key participants to work on the project with the skills that match the responsibilities required as defined in the collaboration specification. This recruitment should ideally be carried out at project organisation level.

Evaluate collaboration solution

This task checks whether the collaboration environment being implemented is in line with the collaboration vision and whether it is likely to achieve the objectives defined in the final collaboration specification. Short-term periodic performance evaluation should carried out to check whether the collaboration environment is working as intended and whether there are any barriers that should be overcome.

Description of the MOCHA process

This process is concerned with the planning and implementation of organisational change resulting from the implementation of collaboration solutions. It follows a similar structure used in the ICE process based on the broad activities define, plan, implement and evaluate.

Initiate organisational change management

This task starts the change management process by identifying the key players, capturing relevant information and defining changes likely to be required as a result of planning and implementing the collaboration solutions:

- *Build the change implementation team*: The need for collaboration is communicated to each participating organisation in order to initiate the formation of the change implementation team, which includes the collaboration champion and other members responsible for the change management. These must be capable of carrying out the responsibilities defined above. Within the CCMT, it is important to have representation from each group that is likely to be affected by the change. Members of the CMT at the project organisation level should also be involved in this process.
- *Communicate initial change*: Initial changes are identified after the preliminary collaboration specifications are defined. It is important to communicate these to employees early in the process to give them adequate time to understand and adjust and hence reduce their resistance to the changes.
- *Capture user and organisational requirements*: This process captures the user and organisational requirements for the collaboration environments to enable the CMT members to make appropriate decisions on collaboration. The collaboration champion and the CCMT members

will then define the detailed changes required to meet the project collaboration needs and the collaboration tools' requisites.

Develop the change vision

This main task defines the need and vision for organisational change and develops an overall strategy to be agreed by each organisation and then communicated to employees:

- *Define the need for organisational change*: Based on the collaboration vision defined earlier, the collaboration champion and CCMT will decide whether there is a need for organisational change and then define its extent and implications.
- *Define change vision/strategy*: The change vision/strategy is then defined; this should be clear enough to be understood by all employees to ensure smooth adoption and implementation.
- *Communicate change vision/strategy*: The above vision/strategy should then be communicated to all employees to ensure good understanding and wide participation and hence guarantee an effective implementation.

Plan organisational change management

This process aims to develop a plan of how the organisational change should be managed:

- *Review current ICT tools and define needs*: The current ICT tools used are established and summarised by the systems manager and CCMT in the form of an ICT audit report for each organisation. These are then considered by the CMT members to identify new technical requirements when planning the collaboration solution.
- *Identify required organisational changes*: The collaboration specification defined at the project organisational level is communicated to each organisation to enable it to establish its impact on employees, organisational processes, existing ICT tools and technologies, and the organisational workflow structures. Taking all these factors and the change vision into account, the CCMT and collaboration champion should identify the organisational changes required to implement the collaboration specification. The output for this process is the list of changes.
- *Plan organisational change*: Once the required organisational changes are identified, the collaboration champion and CCMT should develop a plan for the implementation of change, taking into account the change vision and resources; this will be referred to as the Change Implementation Action Plan in the following sections.

Implement organisational change

This is aimed at finalising the collaboration solution through a pilot implementation, carrying out training, and implementing the solution in the project:

- *Pilot and refine change*: This process carries out a pilot of the collaboration environment implementation and the changes required for this. The changes are implemented according to the Change Implementation Action Plan. Only a sample of users are involved in the pilot; it is recommended to use the CCMT members for this sample since they have already been involved in capturing user and organisational requirements and know what is expected from the collaboration environment. The systems managers should also be involved in this process to cover the technical aspects. The pilot implementation is expected to provide feedback on two areas: the first is related to the change implementation procedures of the organisation, which may be used to modify the change implementation action plan; the second area is related to the collaboration environment implementation. The organisational reflections and feedback on the pilot implementation constitute the main output of this process, which is communicated to the project organisational level to be used to determine any alterations to the collaboration solution.
- *Conduct training*: Training is carried out following the recommendations defined by the CMT. The collaboration champion and systems managers should be involved in this process. This should be considered as a combination of training on the use of the collaboration environment tools and training on the collaboration standards and procedures agreed by the CMT. During the training, some feedback on change implementation is also obtained. This feedback can then be used to alter the change implementation action plan.
- *Implement change*: This process kickstarts the implementation of the final collaboration plan and associated changes fixed after the pilot implementation at the project organisational level. The systems manager, collaboration champion and CCMT should be the main actors for this process. This process enables the start of the collaboration environment usage within the construction project. During this, feedback on change implementation is obtained and used to revise the Change Implementation Action Plan.

Evaluate organisational change

This process is concerned with the evaluation of the adoption of organisational change in the short and long term and facilitates organisational learning during the implementation:

- *Obtain short-term results on collaboration*: A short-term performance evaluation should be carried out within each organisation to check whether the collaboration environment is working efficiently and to identify any bottlenecks or barriers that should be removed. The organisations should also evaluate whether the organisational change is accepted by employees and if there are any alterations to be made in the Change Implementation Action Plan. This process is carried out by the CCMT and collaboration champion using appropriate data collection and performance measurement techniques. The results are documented as the organisational performance evaluation report, which is sent to the project organisational level for a further interorganisational evaluation of the collaboration environment performance.
- *Implement revised collaboration solution*: The modified collaboration solution determined at the project organisational level according to the short-term performance evaluation is implemented in this task.
- *Revise change management procedures*: As previously stated, the organisational performance evaluation focuses on both the collaboration environment and the organisational change. The modifications required in the collaboration solution are determined at the project organisational level whereas the modifications required in change management procedures are carried out in each organisation independently. This process determines the modifications to the change management procedures and is carried out by the collaboration champion and CCMT.
- *Evaluate long-term results*: At the end or close to the end of the project, the long-term implications of the change management are carried out by the collaboration champion and CCMT using appropriate performance measurement techniques. The results of the evaluation are documented as the long-term performance evaluation report.
- *Document organisational learning*: The aim of this process is to document and save strategic data which will be beneficial for future organisational change management attempts in the form of guidelines or recommendations. During any change implementation, organisations go through an implicit and explicit learning process. Best practice, required alterations and corrected mistakes all form lessons learnt during the implementation that should be captured throughout the implementation. The collaboration champion and senior management carry out a final evaluation of these captured lessons at the end of the project to form part of the organisational learning.

Summary

This chapter presented the ICEMOCHA framework developed in order to improve the collaboration in construction projects following both strategic management and organisational change management principles. The model consists of two interdependent models acting at two different levels: ICE and MOCHA. ICE is used at project organisational level and aims at the planning and implementation of effective collaboration environments. MOCHA is introduced at the organisational level and aims at supporting the adoption process in each organisation, focusing on the management of organisational changes required as a result of the introduction of collaboration environments.

Chapter 9

Collaboration implementation in construction

Case studies

Bilge Erdogan, Ozan Koseoglu, Dino Bouchlaghem and Yasemin Nielsen

Introduction

As already highlighted in this book, collaboration is an essential part of construction projects because of its multiorganisational and geographically dispersed project nature. Collaboration tools and systems are commonly used and the industry is constantly seeking new, more efficient and more effective information and communication technology (ICT)-based solutions. Despite emerging technologies offering the construction industry opportunities for computer-supported collaboration environments, the companies adopting these technologies commonly fail to achieve the full benefits from their implementations. This is usually attributed to the priority being given to the technical factors over other aspects related to change management, human and organisational issues, and the roles of the high-level management and end users. Each new information technology (IT) implementation involves some form of change for the organisation and the employees, and is therefore a potential source of resistance and uncertainty unless special attention is paid to managing the related change. This chapter presents the findings from two sets of case studies: the first explored the general issues associated with the implementation of collaboration environments within construction firms (collaboration environment implementation case studies); the second investigated the use and potential benefits of mobile collaboration technologies on construction sites (mobile collaboration technologies case studies).

Collaboration environments implementation case studies

It was highlighted earlier in this book that the main problem in the implementation of collaboration environments is not always related to technical issues but often includes people and organisational factors. In order to obtain an industrial perspective on this, a case study-based investigation of the current collaboration environments (CEs) implementation and collaborative working approaches in construction companies was carried out and aimed to capture information on:

- CE implementation procedures;
- barriers and difficulties in the implementation of CEs and collaborative working procedures;
- the extent to which CE implementations undertaken so far have been successful; and
- thoughts and experiences of industry professionals regarding the transformation of the organisation during the implementation of a new CE.

Background of the case study companies

Nine case studies were conducted involving semi-structured face-to-face interviews with senior managers in construction organisations. Background information on the sample of companies is presented in Table 9.1. The choice of companies was based on the following factors:

- Construction organisations that had been involved in large-scale projects in which many companies collaborated.
- Some of the industrial partners who contributed to the PIECC research project covered in Chapter 3 of this book. These companies were involved in many construction projects in which various types of CEs were implemented.
- Technology providers identified as the main suppliers by the construction companies involved in the study.

Table 9.1 Background of case study companies

Case number	*Company type*	*Job of interviewee*
1	Consultancy	Collaboration consultant
2	Consultancy	Senior consultant
3	Contracting	Corporate service head for quality
4	Contracting	Project collaboration analyst
5	Contracting	Project director
6	Architecture	Associate of practice
7	Architecture	Associate of practice Information manager/document controller
8	Technology	Head of corporate communications
9	Technology	Director in executive management board

Case study results

The qualitative information from the case studies was analysed using a combination of qualitative coding, interpretation and cross-case analysis, whereas the responses to closed questions were analysed quantitatively. Coding techniques were used in order to organise the raw textual data. Coding is the process of identifying and recording one or more discrete passages of text or other data items that cover the same theoretical or descriptive idea (Gibbs, 2002). As part of an analytical process, attaching codes to data and generating concepts enabled the researcher to extract the meaning of the data (Coffey and Atkinson, 1996). Coding is usually a mixture of data reduction and composition used to break up and segment the data into simpler general categories; and to expand and tease out new issues through different levels of interpretation (Coffey and Atkinson, 1996). Following the coding principles, the textual data in each interview transcript were broken down into the main subject categories. For some categories containing complex and complicated information, software for qualitative data analysis was used for the coding.

The analysis of the interview data was then combined with a systems thinking approach to create a causal loop diagram representing the organisational issues when implementing CEs for construction projects. The findings from the case studies are discussed under the following five headings:

1 collaboration technologies implemented;
2 failure rate of CEs to provide the full benefits expected;
3 success criteria for collaboration IT implementations;
4 user involvement during CE implementation; and
5 factors affecting the success of the CE.

Collaboration technologies implemented

The most common CEs implemented by the companies interviewed were project extranets provided by various technology providers. The collaboration technologies acquired and used by the companies interviewed are summarised in Table 9.2.

Failure rate of CEs to provide full benefits expected

Regarding the success rate of the collaboration technologies, it was found that up to 30 per cent of the technologies implemented failed to provide the full benefits expected, whereas Cases 1 and 6 reported an even higher failure rate ranging from 50 to 70 per cent.

Table 9.2 Collaboration technologies implemented

	Collaboration tools used
Case 1	Project extranets (buzzsaw, BIW, Asite, 4Projects) Plan weaver Shared drives Net meeting Videoconferencing Whiteboards 3D modelling
Case 2	SAP Extranets (BIW, Business Collaborator, 4 project, Buildonline) Their own EDM
Case 3	Extranet systems by BIW Information Channel An accounting system called MENTOL Intranet system by Inter-link
Case 4	Extranets by 4Projects (all the time), BIW Project nets by Athena, Sysnet
Case 5	Project extranets (Asite is the principal one, 4Projects) Lotus Notes as EDMS
Arch 1	Project extranets built with buzzsaw
Arch 2	Project extranets by BIW, CADWeb, 4Projects, project net, project web, Buildonline

Success criteria for collaboration IT implementations

When questioned about the criteria used to measure the success of the collaboration implementations and the extent to which these implementations satisfy these criteria, the interviewees reported that they mostly did a perceptual analysis of whether their employees worked better than previously and whether they were more efficient or more productive. Compared with previous projects, some examples of their perceived successes were noticeable decreases in the number of complaints from employees, in the number of requests for information (RFIs) and in rework. However, these indicators are not only related to the success of the CE; they are also affected by many other organisational and project-level factors. Therefore this perceptual analysis does not measure the real success of CE alone. In Cases 4 and 6, the success of CE was assessed by calculating the tangible business benefits in terms of cost and/or time savings through comparisons with other cases where paper-based systems were used. However, these comparisons would measure the efficiency of the collaboration environment as a whole rather than individual tools; for example, documents are sometimes exchanged electronically by email or FTP (file transfer protocol) even when an extranet system is available. Furthermore, the benefit estimated this way would reflect more the effect of the automation of communication than that of the collaboration environment in general.

Companies found it difficult to measure the intangible benefits such as the savings due to reduced rework and RFIs resulting from the use of a specific system. When they needed to quantify these, they either relied on perceived and subjective views or measured the success of the project as a whole instead of the specific benefits of a collaboration tool against a number of criteria or key performance indicators.

The technology provider companies stated that specifying a universal cost saving was difficult and hence they relied on a perceived view for the success of CEs by trying to identify whether companies implementing them were satisfied and whether they planned any future implementations.

The architecture company (Case 6) defined 'ease of transfer of information', 'no repetition of information' and 'ease of communication' as the criteria that define a successful CE implementation. The company measured 'how many times drawings are revised during the project' in order to assess the efficiency of the CE. According to the interviewee, if there is a clean flow of information and a clean flow of communication through the CE, there should not be too many drawings going forward and backward during the process.

None of the perceptual analyses carried out by any of the companies managed to judge the performance accurately; they were mostly subjective and did not include the thoughts of the end users. Unfortunately, these analyses failed to provide results that could be used as feedback for future implementations.

User involvement during the collaboration environment implementation

The CE implementation process is usually divided into a number of different steps. The following nine stages define the CE implementation process:

1 recognising the need for a new system;
2 feasibility analysis;
3 user requirements capture;
4 design of the technical system;
5 planning the adaptation process;
6 choosing the optimum among the adaptation alternatives;
7 testing and evaluation;
8 implementation; and
9 fine tuning.

The interviewees were given a list of users including (1) senior manager, (2) IT manager, (3) construction project manager, (4) external IT specialists, (5) end users and (6) external change agent/consultant, and were asked to identify who were actively involved in the different stages of the implementation process. The technology-providing companies were not asked this question since they would not be able to respond to it based on their own experience. Likewise Case 7 did not answer this question, stating that the company had not actually been involved in the implementation of CEs but participated in the environment via the document controller after everything had been set up by either the client or the contractor. The company was not involved in any of the decisions in the CE implementations other than agreeing on protocols and file formats to be used. None of the employees other than the document controller was actually using the CE. All the work regarding the CE such as uploading or downloading a file to the system was carried out by the assigned document controller.

The results obtained from the six case studies are summarised in Table 9.3. The numbers in the boxes indicate the number of companies stating that the user listed in that column was involved in the CE implementation step shown in the left column of the row. For example, all companies answering this question stated that senior managers and IT managers were involved in the 'feasibility analysis' stage. The number of companies indicating that construction project managers were involved in this stage was three, whereas for the users and external change agents involvement was only reported by one company.

Table 9.3 User involvement in CE implementation steps (total number of companies = 6)

User involvement matrix	*Senior managers*	*IT manager*	*Construction project manager*	*External IT specialists*	*End users*	*External change agent/ consultant*
Recognising the need for a new system	6	3	4	0	3	1
Feasibility analysis	6	6	3	0	1	1
User requirements capture	4	5	4	1	4	0
Design of the technical system	3	6	1	3	1	1
Planning the adaptation process	4	5	2	1	3	1
Choosing the optimum among the adaptation alternatives	5	5	1	0	1	0
Testing and evaluation	4	5	1	1	5	0
Implementation	2	5	3	2	3	0
Fine tuning	3	6	1	2	3	0

As shown in Table 9.3, IT managers are involved in almost all of the stages. In Cases 4 and 6, they were not involved in 'recognising the need for a new system', which was left to the construction project managers. Likewise in Case 3 'choosing the optimum among the adaptation alternatives' was left to the senior managers of the company rather than the IT manager. In both of the consultancy companies (Cases 1 and 2) and in Case 3, the end users were also involved in defining the need for a new system. In Case 6 both senior managers and construction project managers were involved in this task.

In the 'feasibility analysis' stage, those involved were common to all case study companies; these were the IT manager together with senior managers or project managers or both. In Case 2, end users were also involved in this stage.

In the design of the technical system, planning of the adaptation process and choosing the optimum among the adaptation alternatives, the main decision makers were IT managers and senior managers, whereas in the user requirements capture either the construction project managers or the end users were also involved in the decision making.

The case studies showed that the involvement of end users was limited to user requirements capture and the testing and evaluation of the system if the implementation involved any. In all cases, end users were involved mainly at the training stage after the system has been implemented. Most companies have started using a different method of training to improve quality. Instead of training a large number of employees together, they now train them at different levels and have changed the training process from theoretical to practical. When they start on the job, the trainers remain in the company during the adoption stage and help the users. This method initially costs more than the classroom training type but in the long term the costs decrease since the users get acquainted with the system quicker, and hence problems during the adoption stage are solved faster. This new type of training will also result in employees producing better-quality work. On the other hand, the architectural type of companies had difficulties in finding time for training. Most architects working under strict deadlines were reluctant to attend the training sessions because of work pressures. This is one of the reasons for the document controllers having an important role in the use of CEs within the architectural firms.

When the overall results were analysed, it was seen that the IT managers had a very active role in almost all stages of CE implementations. This shows evidence of the commonly encountered criticism of having 'too much focus on IT' in the CE implementations. Furthermore, the results show that the more involved IT specialists are in the implementation, the more focus on IT is observed. On the other hand this situation can be used positively if the IT managers can be influenced to fully consider people and organisational factors during the implementation. If IT managers followed a well-defined change management process in the design and implementation of the CEs, they would play a more decisive role in the adoption.

Factors affecting success of collaboration environment

Various failure reasons were identified by each interviewee and most of these were found to be interrelated when considered together. When the companies were asked to report on their least successful implementations, employee resistance, inconsistency of the contract terms regarding the collaboration tool to be used and insufficient training were mentioned as the main factors that made some implementations less successful than the others. Cultural problems, lack of trust and unsatisfied user requirements were also mentioned by all of the interviewees as reasons for failure:

- *User involvement*: When the companies were questioned on how they implement a new CE, and how they handle the changes occurring in the system, in the organisation and in processes, the responses were mainly limited to *training*. In agreement with previous findings, it was observed that the companies that do not involve the users during the requirements capture stage complained more about user resistance.

  The importance of *top-level commitment* as a success factor was emphasised by most of the interviewees. If a change is to be introduced to an organisation, then it is important that senior managers believe that it is necessary and act accordingly. They will have two roles: first act as collaboration chiefs and manage the implementation, and second ensure that employees in the organisation use the system. If an implementation is left optional, employees will continue to follow their old ways of working. They will not invest additional efforts to get used to a new system. Therefore, senior managers must make clear when and how a new system is to be used in the organisation. On the other hand, especially in Case 5, it was seen that if this push was implemented by coercion, then the employee resistance to change might not be apparent but could be transformed into a 'hidden rage', creating more problems ahead. Imposing the use of a system on users may not achieve positive results, but making them realise the consequences on the project and the company in the long term, together with some managed enforcement of what is expected, will ensure better participation.

  The technology-providing companies in Cases 8 and 9 indicated that when more than one senior manager was in charge of managing collaboration, and had conflicting opinions, the implementation was adversely affected and this sometimes resulted in long delays while the conflicts were resolved.

  The case studies showed that there were no formal ways of obtaining end user *feedback* throughout the implementation in any of the organisations. However, in some of the organisations, if the implementation included a testing or validation stage, the end users were involved.

- *User resistance to change*: *Early user involvement*, *user-friendly interfaces* and *training* were identified as the main factors that can attenuate user resistance to change.
- *User requirements capture*: It was found that the case study companies did not have formal methods for capturing user requirements. Further, only three companies used a sample of end users to obtain feedback after the CE was implemented.
- *Planning and project management*: *Lack of agreement between parties* is found to be the main cause of failure in this area. It is necessary for the collaborating organisations to *agree on the common formats, types and conventions* for the information exchange before the CE is set up, to provide consistency and avoid possible conflicts. *Incompatibility of the processes*, lack of *contract clauses* regarding the collaboration use and *lack of clear guidance and prospectus* were found as other factors affecting the success of CEs.
- *Technical factors*: Interoperability problems, IT incompatibility of the collaboration environment, unfriendly user interfaces, low speed of transfer and data security problems were mentioned as the main failure reasons related to technical factors.
- *Buy-in from all parties*: All of the interviewees stressed the importance of the collaboration tools being used by all parties for the success of the whole project and highlighted that it should be ensured either by mutual agreement or by inclusion as contract terms. The importance of contractual arrangements regarding the CE used for external communication was especially emphasised by the case study companies. The contract should be binding for all companies participating in the project to make sure that there are consistent procedures for the use of the systems.
- *Trust factors*: Enabling trust between the collaborating parties and the security of data were mentioned by all interviewees. It is important to ensure that the data on the system can be accessed only by the appropriate parties. Furthermore, the transparency of the system should be adjusted to the extent that each private organisational datum is protected. This was considered as an important factor especially by contracting companies since they did not want to lose their bargaining power and their benefits as a result of claims for additional work to be carried out. The case studies revealed that there would be fewer concerns in this area if the type of information to be shared and the mechanisms are well defined at the outset. The architecture company in Case 7 thought that the trust issue could be solved if the CE were implemented and led by a third party, whereas the other architecture company (Case 6) suggested that the CE should be led by the architecture companies since they are involved in the project from the very beginning and remain until the end.

In summary, the success of collaboration is found to be affected by a number of different factors related to organisation, people or technical issues. When the effects of these factors were investigated one by one, based on the thoughts of the interviewees, it was found necessary to categorise them into two groups: factors affecting the collaboration at the organisation level and factors affecting the collaboration at the project organisation level. These factors are shown in Figure 9.1.

The interdependency between some of the factors shown in Figure 9.1 was specifically mentioned by the interviewees. Further relationships were revealed after the detailed analysis of the case studies. These relationships were interpreted using a 'systems thinking' approach and a causal loop diagram (CLD) was developed in order to represent the organisational dynamics during the introduction of a new CE to an organisation. The CLD and the steps followed to develop it are explained in the next section.

Systems thinking

Systems thinking is a method used to enhance learning in complex systems and is fundamentally interdisciplinary. It is based on the ideas: 'You cannot just do one thing' and 'Everything is connected to everything

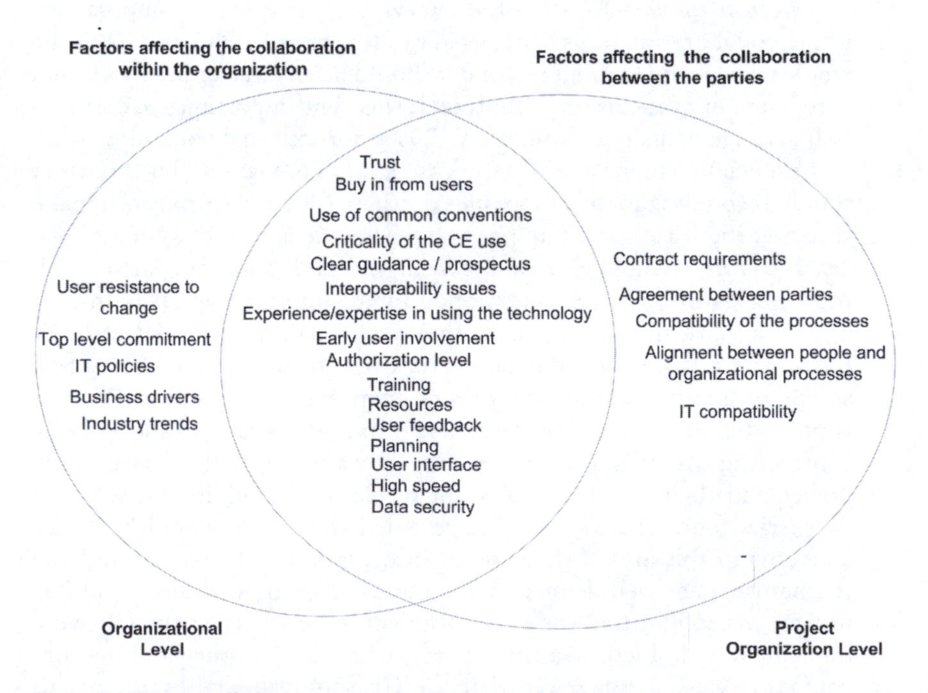

Figure 9.1 Factors affecting success of collaboration at organisation and at project organisation level.

else' (Sterman, 2000: 4). According to systems thinking there are no side effects; there are just effects and feedbacks. The feedback structure of systems is represented by CLDs which are considered to be excellent for (Sterman, 2000):

- quickly capturing the hypothesis about the causes of the dynamics;
- eliciting and capturing mental models of individuals or teams; and
- communicating important feedbacks that are believed to be responsible for a problem.

Causal loop diagrams provide a significant level of assistance to thinking by introducing circular causality and providing a medium by which people can externalise mental models and assumptions and enrich these by sharing them (Wolstenholme, 1999). The CLD elements and their descriptions are presented in Table 9.4.

The main organisational issues for successful CE implementation derived from the analysis and interpretation of the case studies are listed below:

- criticality of the CE implementation for the project success;
- binding clauses in the contract regarding the use of CE;
- agreement between parties on the use of a CE;
- trust factor: between the organisations and in the system;
- security of organisational data;
- top-level commitment and collaboration champion role;
- user resistance to change;
- early user involvement;
- user-friendly interface;
- training;
- consistency of data format, types and conventions between organisations;

Table 9.4 CLD elements

Notation name	*Explanation*
Positive link	All else equal; if $X\uparrow \rightarrow Y\uparrow$ If $X\downarrow \rightarrow Y\downarrow$ (ÔY/ôx>0)
Negative link	All else equal; if $X\uparrow \rightarrow Y\uparrow$ If $X\downarrow \rightarrow Y\downarrow$ (ÔY/c9x<0)
Positive/reinforcing loop	The loop starts with an increase and ends with an increase (or vice versa)
Negative/balancing loop	The loop starts with an increase but ends with a decrease (or vice versa)

- use of common conventions; and
- efficiency of the CE.

These factors are all interlinked; a change in one of the factors influences the others in a positive or negative way. The relationship between two factors is investigated independently from the rest and is represented by a negative or a positive link. For example, top-level commitment is found to have dependencies with four factors as follows:

1. If the use of CE is critical to the success of the project, the top-level commitment will increase to ensure the success of the CE and the construction project. If the use of CE is less critical, then the top-level commitment will decrease. A positive link is used to represent this directly proportional relationship between these two factors.
2. If there is a legally binding statement in the contract, the top-level managers will be more committed to make sure that the organisation will not fail to meet its legal responsibilities. If the statement in the contract is less binding, then the top-level commitment will decrease. This relationship is shown by a positive link in the CLD.
3. When the top-level commitment increases in an organisation, the user resistance will be less. As discussed in the case study results section, the top-level commitment is expected to be a combination of a collaboration champion role and an enforcement role. Employees will be encouraged to use the CE by the collaboration champion while at the same time being aware of the negative personal consequences if they do not follow the collaboration method agreed. Top-level managers should make sure that the collaboration method agreed by all parties at the beginning of the project is followed and implemented within their own organisation. Since a change in the top-level commitment drives the user resistance in the opposite direction, the relationship is shown by a negative link in the CLD. Not only is the user resistance affected by the top-level commitment; top-level commitment is also affected by the user resistance. When the user resistance increases, top-level management will be more involved to solve the problems and to remove the barriers to success. When the resistance decreases, top-level management will be more relaxed and will start to be less committed. Therefore the feedback from the user resistance factor to the top-level commitment is represented by a positive link due to the directly proportional relationship. The negative link from top-level commitment to user resistance and the positive feedback forms a balancing loop and is therefore shown with a 'B' sign in the diagram.
4. If the top-level commitment increases, the use of CE will increase. Likewise, a decrease in the top-level commitment will be reflected in a decrease in the use of CE. Contrariwise, the relationship between the use of CE and the top-level commitment is represented by a negative

link. Top-level commitment and the use of CE relationship will form a balanced loop because of the positive link from top-level commitment to the use of CE and the negative feedback link from the use of CE to the top-level commitment.

The same approach was applied to the rest of the factors in order to understand the organisational dynamics during the CE implementation. Based on the relationships and dependencies found in the case studies, the CLD in Figure 9.2 was developed. As seen from the figure, CLD involves six balancing loops and one reinforcing loop as follows:

1 balanced loop of top-level commitment – use of CE – top-level commitment;
2 balanced loop of top-level commitment – user resistance – top-level commitment;

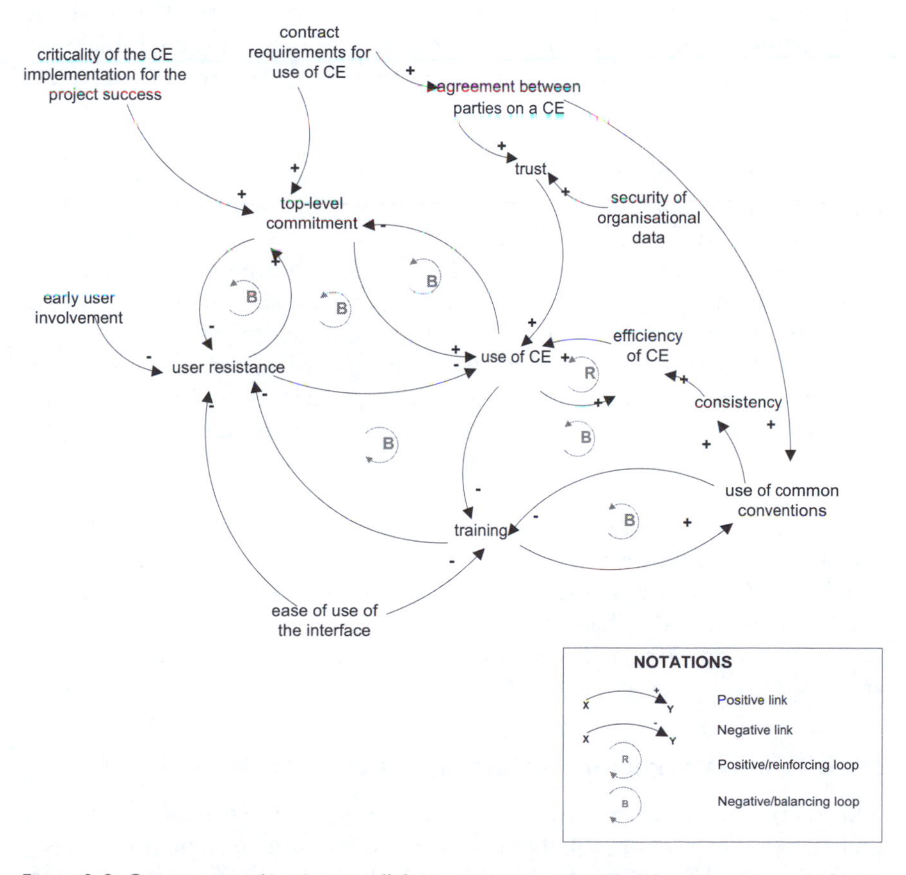

Figure 9.2 Organisational issues in collaboration environment implementation: causal loop diagram.

3 balanced loop of top-level commitment – user resistance – use of CE – top-level commitment;
4 balanced loop of training – user resistance – use of CE – training;
5 balanced loop of training – use of common conventions – training;
6 balanced loop of training – use of common conventions – consistency – efficiency of CE – use of CE – training; and
7 reinforcing loop of efficiency of CE – use of CE – efficiency of CE.

Looking at the diagram, it is seen that the use of CE is either directly or indirectly linked to each of the factors in the loops, or feeds into the other factors in the loops. A positive link has been assigned between the contract requirements for the use of CE and top-level commitment. The existence of a legally binding statement in the contract also influences the agreement between the parties on a CE through a positive link. When there is an agreement between the parties on a CE, the trust between the parties is enhanced and strengthened and also the use of common conventions is facilitated. The trust factor in the CLD includes trust in the CE as well as the trust between the collaborating parties. The trust in the CE increases when the security of organisational data increases. The trust factor has a direct impact on the use of CE. Therefore when the collaborating parties develop trust in each other and in the CE, the use of CE increases.

The training has an impact on two factors: user resistance and use of common conventions. Training reduces the user resistance if it is carried out appropriately and increases the use of common conventions. The training and use of common conventions are linked to each other with a balancing loop due to the positive link from training to the use of common conventions and the negative feedback link in the opposite direction.

Other than this feedback link from the use of common conventions, training is affected by the ease of use of the interface and the use of CE factors. If the interface of the CE is user-friendly, then the amount of training required will be less. A user-friendly interface will also reduce the user resistance.

The use of common conventions will increase the consistency in the CE, which will increase its efficiency. The more efficient the CE is, the more it will be used in the organisations, and the more it is used, the more efficient the CE will become. When the use of CE increases in the organisation, then the need for training will decrease since people will not need any more training.

Mobile collaboration technologies case studies

This section presents the case studies on the implementation of information technologies to support collaborative working within construction projects carried out with two major contractors. The case studies were also supported by a wireless network provider and a software developer.

The first case study was carried out with the collaborative working and project modelling manager in the first contracting form. The aim was to identify barriers and potential solutions for the implementation of mobile technologies to support collaboration between design and construction teams. The use of four-dimensional (4D) modelling and wireless networks was the key feature for the implementing mobile collaboration on the construction site.

The second case study aimed to review current practice in the planning and implementation of mobile technologies within a typical construction business environment. The results revealed that the case study company has an ambitious vision for implementing such technologies in various business units.

Case Study I

This case study was based on a school project, part of the Building Schools for the Future (BSF) government-led programme in the UK. BSF is the biggest single government investment for the improvement of school buildings for over 50 years. The aim was to rebuild or renew every secondary school in England over a 10- to 15-year period (National Audit Office, 2009). BSF brings together significant investment in buildings and in ICT (information and communications technology) to support the government's educational reform agenda (National Audit Office, 2009).

The school project was selected because it made use of the Avanti Collaborative Working Procedures (http://www.constructingexcellence.org.uk/ceavanti/) while developing the three-dimensional (3D) computer-aided design (CAD) model and publishing it on the project extranet (4Projects, http://www.4projects.com/). All project stakeholders, including the clients, had access to the project extranet and were able to provide feedback on the design development. The architectural team stored the 3D model files in the standard AutoCAD format (DWG) on the project extranet. The structural engineers could download the architects' 3D model files and create their own 3D structural drawings. This resulted in two sets of 3D drawing files, architectural and structural. The contractor then checked the coordination of the 3D model in Navisworks Jetstream (http://usa.autodesk.com/navisworks/), an application that combines architectural and engineering design information in a single virtual model; this was also stored within the project extranet site. As a result, both the client and the construction project team were able to download the Navisworks NWD file for review.

Although the project stakeholders could share, download and view the coordinated 3D model files during the design development, the majority of engineers on site still used paper-based drawings with limited access to 3D models whilst monitoring construction progress. The planning and implementation of the mobile visualisation approach presented in Chapter 5 of this book addresses this collaboration gap between the design and construction site teams and proposes a combined mobile digital collaboration model to improve the existing working methods.

This case study focused on the application of the mobile 2D (two-dimensional)/3D and 4D collaboration scenarios presented in Chapter 5. The objectives were to:

- develop a 4D model for the BSF project using the design team's 3D model files;
- identify technical and non-technical problems that occur during the development of the 4D models;
- demonstrate how mobile visualisation applications can be applied to real construction projects for design–construction collaboration.

4D model development

To commence the development and implementation of the 4D model the source data had to be made available in the following formats:

- 2D CAD model file in Design Web Format (DWF);
- 3D CAD model file in DWF;
- project programme data from MS Project or Asta Power Project in XML (extensible markup language) format.

These requirements were specified by Synchro, a 4D modelling software and real-time collaboration platform used in the research (http://www.synchroltd.com/). Synchro Professional version 3.1 was used during the development of the 4D model on a server that provided real-time web-based collaboration so that any changes made by one project member could be broadcast immediately to all other participants. The process of developing the 4D model for the BSF project and the problems that were experienced are summarised below:

- The contractor provided and managed access to the project extranet hosted by 4Projects.
- 3D and 2D CAD model files (DWG format) produced by the architects and engineers were downloaded from the 4Projects extranet.
- The coordinated 3D project model produced by the contractor in NWD format (Navisworks) was examined in order to identify which files were used to develop the 3D model and how they are linked to each other.
- As another 3D and 4D modelling platform, Navisworks does not allow the project team members to access the 4D model in real time though a web-based collaboration platform. Therefore it was used only for displaying and examining the existing 3D project model.

- The project schedule exported from Primavera (http://www.ifsworld.com/) as a PRX (project reporter export format) file was provided by the project team. The Synchro project modelling team converted this file into the XML format and imported it into Synchro Professional version 3.1.
- The first attempt made to integrate the 3D model with the project schedule was to export the Navisworks 3D model as a DWF file and directly import it into Synchro. However, the software operation became too slow because of the large number of references. In the upcoming version of Synchro, there will be a new improvement which provides full access to the DWF format information with optimised data storage. There were also problems with the 3D model in Navisworks; this was assembled using disconnected components containing a large number of 2D plans.
- The solution developed through collaboration with the Synchro 4D modelling team used the original model files (those produced by engineering and architectural teams in DWG format) in order to create a new 3D project model more suitable for use within the Synchro software by switching off the 2D CAD grid layers.
- As a result of all this process the 3D model and project schedule were imported and combined into the Synchro software, whereby users were able to display the new project model through the shared workgroup within Synchro. This allowed end users to access the project 4D model located on the web server.
- The Synchro model is made up of different views which display different project information including the companies involved in the project, the 3D model, material resources, human resources, risk information, Gantt chart, tasks and task properties (general, costs, assigned resources, resource utilisation, etc.).
- One of the main objectives in this case study was to develop a fully integrated (3D plus tasks information) 4D model for the BSF project. A fully integrated 4D model means assigning every object (material resources) in the resources list to the related task. Assigning resources to tasks enables users to monitor resource utilisation and display the 3D model from the start of the project to the end in a sequence using the project's scheduling information. However, in the first instance there were no direct linkages between the detailed programme and the objects in the 3D model. The detailed programme included 160 tasks (activities) which were not specifically linked to any part of the 3D model. The project schedule had to be revised and simplified in order to develop a 4D model which better represents the construction sequence.
- Access was gained to a high-level program which presents the project in three main phases: construction of the main teaching block, construction of the sports hall and refurbishment works. Forty tasks were then

identified (with the help of the contractor's project modelling team) in order to make the schedule compatible with the 3D CAD objects in the project model. A new schedule was then developed using these tasks. The last phase of the development of the 4D model was to link these tasks to the 3D objects in the Synchro platform and update the model on the web server.

Findings from Case Study I

During the course of this case study some interested findings were captured during the project team meetings:

- Construction projects are making use of visualisation technologies to improve the design–construction collaboration practices. It was observed that a project extranet with visualisation technologies and collaborative working procedures were used by the BSF project team. However, some of the important components missing were the fully integrated 4D project model and real-time mobile communication infrastructure on the construction site.
- 4D modelling can be a lengthy process if the project plan is not appropriately developed and compatible with the 3D objects in the CAD model. This is generally caused by the fact that 2D drawings and 3D models are produced by designers whereas schedules are separately developed by contractors who lack knowledge on the 3D modelling process.
- The meetings attended during the BSF case study and semi-structured interviews conducted with the design manager, CAD manager, technology consultant and principal engineer involved in the project revealed some key issues about the implementation of collaboration technologies on live construction projects including potential barriers and benefits, solutions to overcome barriers and specific use in off-site construction methods. The key findings from the meetings and interviews are summarised as follows:

BARRIERS

- People are generally reluctant to change their traditional methods of working.
- Construction organisations involved in some projects do not always fully embrace the use of ICTs because of fear that they will not work properly and hence will be a wasted investment.
- There may be poor wireless networking and mobile telecommunication performance at remote construction sites.
- Construction project teams do not always follow standard procedures for sharing, developing and accessing project design information and documents.

POSSIBLE SOLUTIONS

- Educate end users in projects about the potential of mobile and wireless technologies in facilitating on-site collaboration.
- Though exemplar projects, demonstrate how the use of mobile technologies for collaboration can reduce project costs and manage the risk of cost and time overruns.
- Collaborate with wireless and mobile network technology providers to provide 3G (third-generation), Wi-Fi and WiMAX (worldwide interoperability for microwave access) technologies for construction sites.
- Develop clear and robust procedures for the implementation of mobile collaboration technologies to avoid problems caused by the incompatibility of data formats and modelling platforms.

BENEFITS

- Mobile collaboration enables all stakeholders to access an integrated 4D model at any point in time within the project programme and to identify any problems well in advance of them happening on site.
- Using mobile technologies improves communication and collaboration on site as the user can view the 4D model; access published drawings, videos and animations; and upload videos from the site.
- Designers generally produce designs and then move on to other projects before construction starts. The implementation of mobile technologies can offer a virtual collaborative working platform where designers can provide their coordinated design information faster and more accurately, hence reducing the amount of time spent on rework.
- Finally, implementing mobile technologies on site reduces the risk of buildability problems occurring, and avoids unforeseen additional costs and time delays to projects, by enabling the project team to collaborate in real time.

OFF-SITE CONSTRUCTION

A specific construction type which can particularly benefit from these technologies is off-site production systems. In these methods, fabricators supply 3D models in different data formats not always compatible with those used in the construction industry. Mobile technologies can make it easier for site operatives to install the offsite-fabricated components if the 3D drawing files are made available for the development of a 4D model.

Case Study 2

The aim of this case study was to evaluate the planning and implementation of mobile collaboration technologies within a live construction project and business environment. To achieve this aim the objectives were to:

- capture key information about the design development process, information communication and collaboration;
- map the process of design information flow between design and construction site teams;
- follow the stages for the planning and implementation on a pilot study to identify potential users, current ICT systems and technology requirements, and cost benefits.

The case study began with the collection of the necessary data organised in two stages.

Stage 1

- Collect key information about the design development process, communication and collaboration throughout the project lifecycle (including data formats, modelling platforms and collaboration technologies used).
- Model the existing processes used by the project stakeholders for design development and collaboration.
- Conduct semi-structured interviews with the personnel responsible for design development and project collaboration to identify potential benefits, barriers, solutions and potential savings in terms of time and cost as a result of the use of mobile collaboration technologies.

Stage 2

- Arrange a construction site visit to model the process of design information communication between design and construction site teams. This was also used to identify existing collaboration and communication tools used for the following:
  - information sharing and access;
  - monitoring of construction progress against the programme and design specifications;
  - decision making to respond to unexpected problems.
- Conduct semi-structured interviews with key site personnel to capture their views on the use of mobile technologies on-site and to explore potential benefits, barriers, solutions and savings in terms of time and costs.

Data collection and findings

This case study included semi-structured interviews, observations and field work in addition to the collection of information from company reports. In the first week of the research semi-structured interviews were held with various experts at the company's head office. The main aim of these interviews was to identify opportunities for or barriers to the following implementation stage and to understand the company's working practices and processes in various areas such as production control, visualisation technologies and project collaboration. The interviewees' profiles and their years of experience are presented in Table 9.5.

During the interviews, interviewees were also asked to share their personal experience in the organisation and to provide any further documented information on production control, use of visualisation, general and construction site (project) collaboration strategies. Data collected from interviews, observations and written documents are all combined as case study findings and presented in the following sections.

VISION

The company's overarching vision is to identify and adopt new technologies that can make the construction processes more efficient and intelligent. A business improvement team was formed to provide support for a high-profile project and then rolled out to cover the whole business. The team uses a collaboration toolkit that includes digital (virtual) prototyping, process mapping, project flow, logistics and supply chains, and continuous improvement. The company controls the production using the toolkit and tries to

Table 9.5 Head office interviewee profiles

Main role in organisation	*Experience (years)*
Business improvement manager	13
Business improvement manager	28
Business improvement manager	12
IT manager	23
Regional IT manager	25
Digital prototyping leader	17
Digital prototyping	16
Management trainee	3
3D modeller	22
3D modeller	14
3D modeller	22
Senior document controller	3
Head of knowledge management	unknown
Project extranet controller	unknown
Collaboration consultant	19

integrate other innovative technologies such as 4D and 5D modelling, laser scanning, and mobile and wireless technologies in order to improve its existing processes for production control.

DIGITAL (VIRTUAL) PROTOTYPING

Digital prototyping is defined by the company as 'the use of 2D and 3D CAD tools to visualise the construction of a project "virtually" before it is built physically'. It is recognised as providing various business benefits including:

- establishing the correct spatial 'fit' of assemblies and components;
- rehearsing and optimising the sequence;
- assisting in the identification and quantification of materials.

It is common practice for project teams in the company to agree on a data exchange format which can be read by various software systems used by other participants. Preferred file formats are specified by the company when exchanging design information. These are:

- Bentley Microstation – DGN format;
- AutoCAD ADT and CAD-Duct – DWG format;
- UGS Solid Edge.

Agreeing a suitable common data format for the exchange of information is a key issue for collaborative design. Portable document format (PDF) is a preferred option for the publication of 2D drawings (DWF is also commonly used) whereas DGN and DWG are the preferred formats for model files in 2D and 3D. The company adopts a standard procedure for the exchange of design information with other parties involved in projects. It starts by setting an origin and orientation for the model of the project whereby every author generates model files using their own but suitably compatible protocols so that the different parts of the model are combined within the same grid using a collaboration tool that includes a 3D review option.

The digital prototyping team based in the head office assists various projects in 3D, 4D and even 5D modelling and provides support and training in its development and use when 3D modelling is used in the project. The team's role is to promote the use of digital prototyping throughout the project lifecycle. When project model files and drawings are received in 2D form the digital prototyping team and other internal technical support groups create project-specific 3D visualisation models which can be used for client briefing, clash detection, construction sequencing, animation and other purposes. These models are not detailed enough to represent every aspect of the detailed design for all projects, as that is both unnecessary and costly. In some projects, 3D models are updated according to the changes that have occurred in the design information during the construction stage. In the past, 3D models were mainly used during the meetings with clients, architects, mechanical and electrical engineers and structural engineers to present the expected end product before the construction starts on site and identify any clashes between systems or inconsistencies with the project brief. More recently the company started experimenting with the use of these models on site. Some of the benefits realised when using 3D models on construction sites, based on the case study company's projects, are as follows:

- improvements in the monitoring process used on site during the monthly contract reviews;
- phasing and programming of the works before the actual construction starts;
- improving health and safety with the help of visual method statements (animations) generated from 3D models showing the systematic construction process and revealing any possible problems;
- improving the quality of the product delivered to the client.

For the future the company aims to become more involved at the design stage of projects as a constructor and manufacturer. This will enable it to address buildability issues using digital prototyping during the early stages of the project's lifecycle and explore the use of the company's own off-site-manufactured standard components. Moreover, the company predicts that, in future, most designers will be able to produce 3D models of their designs as part of standard contracts, making digital prototyping become easier and common practice.

PROCESS MAPPING AND PRODUCTION CONTROL

In one of the best-case examples for the use of a 3D prototyping, process mapping and project-flow toolkit on a site, the case study company received the whole 3D information from the designer in a block model. The virtual prototyping and the business improvement teams divided the model and the attached main programme of work into production phases to make the project information understandable and easy to use by the site teams and to be able to monitor the actual production on site.

These 3D model files are converted into 2D drawings for the site team using Solid Edge software (http://www.plm.automation.siemens.com/en_gb/), which is commonly used in the aerospace and manufacturing industries. Moreover, visual method statements (animations) can be generated using the 3D model files in order to improve site safety and make site teams better understand the construction phases. In a recent high-profile project, the site team worked closely with site-based designers and production planners. They ran regular daily meetings based around the 3D representation of the job and production plans.

Once the construction sequence is assimilated by the site teams, they can develop their production plan. Production control tools are then used to manage the day-to-day completion. After the process-mapping stage, the project flow method is used to monitor the construction processes and allows the project team to capture reliable information on the actual production on site.

GENERAL COLLABORATION ISSUES

In recognition that improving collaboration in projects is also about people's behaviour and working methods, the company formed a collaboration centre facilitated by the use of various visualisation technologies in order to bring project teams together to review design information at the various stages of the project's lifecycle. The collaboration centre was introduced within the main offices to facilitate collaborative behaviour between all stakeholders. It has the following facilities which enable project teams to review 3D prototypes and animations from projects, identify clashes and solve buildability issues in a collaborative manner:

- Digital Prototyping Suite with 5.3 m by 2.4 m screen;
- Digital Prototyping Room with 2.4 m by 1.8 m screen;
- 20 networked desks with telephones in the central atrium;
- board room with audiovisual facility, DVD player, conference call capability etc.

A collaboration extranet system is used for various projects and forms a major part of the collaboration strategy in terms of sharing drawings with project members. It is used by designers and architects during the construction stage of the project. From the contractor's point of view, the use of project extranets is very similar to using a postbox throughout the project's lifecycle. This reflects the culture and attitude of the construction industry towards collaboration technologies. The most important benefit of using the extranet within the company's projects, in addition to improving communication and collaboration, is to have better control of project records and documents.

IGATE (http://www.igate.com/) is used as an internal collaboration tool which enables everyone to communicate and publish best practices and innovations captured from various projects. IGATE forms a part of the knowledge management strategy and the future aim is to increase its rate of use to share more formal reports about innovative developments within the company.

As real-time collaboration tools, the Windows Messenger instant messaging program and online web conferencing are used in some projects. There are also some fixed videoconferencing facilities at all regional offices but there is no plans to invest in mobile videoconferencing technologies in order to collaborate with construction sites.

There is a lot of effort being dedicated to implementing 4D and 5D solutions in construction projects. If and when the digital prototyping techniques are fully operational in a collaborative environment, the company plans to use them to monitor cost and schedules virtually and abandon the existing reporting system. This will enable teams to make better collaborative decisions in real time. There are also some trials being conducted for mobile collaboration technologies within the various business units with the aim of bringing design and construction site teams together within a multidimensional modelling environment.

COLLABORATION ON SITE

Some of the company's construction sites are equipped with wireless access points but provide coverage only within site cabins. On the other hand, the company provides 3G data cards to be used with laptops. This provides internet access for users but with limited download speed.

The use of mobile devices such as personal digital assistants (PDAs) and tablet PCs is currently limited. PDAs, for example, are used asynchronously for on-site inspections without wireless access. Another on-site application, which is designed for labourers, is mobile time-keeping software integrated with global positioning system (GPS) technology and a camera on a PDA device. This application captures information and sends it to the project website so that the main office can monitor the positioning of the site teams. The mobile time-keeping application is mainly used by mobile workers who do not stay in a fixed location. There is also a pilot project on the use of digital pen and paper to mark up documents on site and send them to a CAD workstation in the form of electronic information.

Pilot project implementation

The Radical Innovation Group within the company on which Case Study 2 was based supports projects and other business units for the implementation of new technologies, methods and innovative solutions. One of the group's internal initiatives aimed to improve the existing reporting system used in projects in order to better support decision making and access real-time progress information during the project's lifecycle. A landmark project in London was identified as a suitable test environment for the initiative and therefore selected for a pilot project in this case study.

Construction site findings

Semi-structured interviews were held with various staff at the construction site office. The main aim of these interviews was to identify drivers and barriers associated with the implementation of mobile technologies, and capture attitudes towards the use of these technologies. The interviewees' job titles and their years of experience are presented in Table 9.6.

The project site engineers still communicate and share design information using paper-based drawings downloaded from the project extranet. They do not usually make use of all the facilities offered by the web-based system including drawing mark-up and circulation on a digital platform. Engineers usually respond to request for information by marking up the drawing on paper and then uploading a scanned copy onto the extranet or emailing it as an attachment to relevant project members.

Progress on site is monitored in daily and weekly meetings. Project flow is used as a production control tool whereby progress information is updated during the meetings. This provides various statistical data related to the construction process. Project flow is a useful tool for the management of short-term workload but additional work is required to link it back to the main programme. The 4D models have the potential to fill the gap between the production control system and the main programme. Furthermore, if the 4D modelling platforms are upgraded to process and store detailed information about the programme tasks, then keeping separate records within a production control tool is no longer necessary.

The company's digital prototyping team made the 4D modelling software (Synchro) available on the construction site to enable engineers to better visualise the construction schedule and allow the project team to monitor progress in real time. Since 4D models are not generally created during the early stages of projects, additional effort was required to link the main programme to the 3D building model. On the other hand, designers do not provide detailed 3D models at the early stages of the design process, therefore construction planners do not usually have the opportunity to use a 3D

Table 9.6 Construction site office interviewee profiles

Main role on site	*Experience (years)*
Project manager	28
Construction manager	19
Section manager	8
Project engineer	9
Planner	3

model when developing the main project programme. Currently, site teams use two different systems; short-term project flow (production control) and the overall project programme. The implementation of 4D modelling could potentially bring the two systems together and enable site teams to have a quicker and better understanding of the construction sequence. Another important benefit of using 4D models on site is the improvement they can bring to the monthly progress reporting. On construction sites, planners are responsible for recording progress according to what is actually built so that cost managers can monitor the cost of resources on a weekly basis. Considerable efforts are spent on the development of monthly progress reports using systems based on historical rather than actual data. Therefore, the case study company investigated the use of digital prototyping to make the preparation of monthly progress reports on site quicker and easier.

The key issue with the deployment of mobile technologies on site is to carefully understand user requirements in terms of applications and devices. Voice communication through mobile phone networks is the only method available to the project team for remote collaboration. If managers and engineers working on site were equipped with other mobile devices such as Pocket PCs with integrated cameras and voice-recording capabilities, they would save considerable amounts of time while creating their daily reports and snagging lists. The interviewees confirmed that the use of Pocket PCs for the recording and conversion of audio data into text material, capturing photos and video clips, and accessing wireless networks would improve site communication and collaboration, and would potentially achieve cost savings during the execution of projects. Furthermore, the use of tablet PCs to display 2D drawing files, 3D model files, 4D representations of the construction schedule, and visual method statements on site would improve the efficiency of communication between design and construction teams, and shorten the time needed to access information on site. However, users have some concerns about being able to read drawings on a small digital screen, network speed and the robustness of the devices in a rough site environment. Another important obstacle in the way of implementing mobile technologies on site is related to users' resistance and lack of knowledge. The next sections present the planning of the wireless network layout for the project; and identification of the potential users, technology requirements and the necessary investment for the implementation of mobile visualisation at the project.

WIRELESS NETWORK PLANNING

Project information including site dimensions, the height of various buildings and construction sequence was captured from the 4D model. The initial plan would be to bring a 2 Mbps broadband option into the site office and mount an access point (AP). Initially, the site is clear and only a few APs are needed. Depending on the number of users, two or three APs would be enough to cover the entire site; the exact locations are not critical and nodes can be moved as required. However, as construction proceeds, the erection of buildings will create radio shadows which will necessitate some fine-tuning of the system, for example mounting APs on top of cranes, raising them above the level of the buildings to facilitate cross-site coverage.

As the buildings near completion, the height may hamper good propagation depending on the exact mounting of the nodes. For example, it could be desirable to mount an AP on top of a crane to provide good rooftop coverage. To achieve this, a pair of wireless bridges would be required with one mounted on the site office and the other close to the AP on top of the crane with the antenna panels directed at each other for optimum performance. In addition, floors that require coverage might need to have APs, about one for every three floors, and could be moved to suit the coverage required. The exact requirements and coverage heavily depend on the materials used for the construction of the buildings.

COST BENEFITS

A recent survey conducted by YouGov in association with T-Mobile on the use of mobile technology on construction sites involved 380 construction professionals and revealed the following (T-Mobile, 2006):

- 40 per cent of construction companies do not provide staff with any form of mobile devices including phones;
- 19 per cent believe that their employers are failing to exploit fully the potential of mobile technology;
- mobile devices can save construction staff an average of 34 minutes per day.

The results of this survey were used to estimate the potential time saving and the equivalent monetary value of the associated indirect cost savings. According to the above survey, it is assumed that using mobile devices would save construction staff an average of half an hour per workday. Calculating the time saved by site staff during each year of the case study project resulted in the estimated figures shown in Table 9.7 (based on 300 working days a year and an average time saving of 30 minutes per day).

Table 9.7 Total time saving throughout the project lifecycle

Year	*Number of users on site*	*Yearly time saving (man-hours)*
Year 1	10	1500
Year 2	30	4500
Year 3	30	4500

In monetary terms, the above time savings resulting from the implementation of mobile technology would constitute a return on investment of about 18 per cent. This does not include direct savings in terms of reduced phone costs, travel, delays and so on.

Conclusion from the case studies

Conclusions from the collaboration environments implementation case studies

- All companies were found to be failing to achieve the full benefits of CE implementations because of their underestimation (or ignorance) of the people and organisational issues.
- For the success of the whole project, the collaboration tools should be used by all parties in a project. The criticality of the tool for the success of the project will play an important role in the extent of use. Second, the contract terms regarding the CE to be used should be clear and binding for all parties to obtain commitment and consistency.
- The transparency of the data in the CE should be arranged carefully to prevent any possible resistance by the parties to use the system.
- The common formats, types and conventions for the information exchange should be agreed before the CE is set up.
- User interfaces should be user-friendly.
- Senior management commitment by means of a 'collaboration champion' accessible to the end users should be balanced with enforcement actions.
- The results have shown that there are strong links between the success of the CE implementations and user involvement, and between employee resistance and user needs capture. It has been shown that the more and the earlier the users are involved in the design and implementation of the CEs, the better the user requirement will be captured and the less resistance will occur.
- Employee resistance should be dealt with appropriately depending on the sources and extent of resistance.
- The changes brought about by the CEs should be managed at the organisational level.

Conclusions from mobile collaboration technologies case studies

- The realisation of the aim of mobile visualisation for on-site collaboration has become possible owing to the increase of computational power and speed of wireless technologies.

- The integration of mobile, wireless and visualisation technologies on site has significant potential in creating a collaborative working environment for designers and construction site teams. The research results revealed that the use of these technologies on site could also improve health and safety of the workforce carrying out construction tasks.
- The implementation of mobile, wireless and visualisation technologies should be planned carefully, taking into consideration technology, process and human-related requirements. Apart from the technological issues, construction professionals need to adapt to the concept of virtual collaboration.
- Visualisation-based mobile collaboration on site can be more effective than paper-based methods as it supports project members in coordinating work, understanding the construction sequence and monitoring progress.
- The deployment of mobile and visualisation technologies on site is highly dependent on the clients, contracts and the perception and willingness of end users. In traditional procurement methods, contractors are not involved in the early stages of projects, when collaboration tools and technologies are usually identified and implemented. Involving contractors early in the design process can improve the quality of the final product by tackling buildability issues.
- The lack of mobile technologies adoption by the end users can be easily overcome by raising awareness and the introduction of training and personal development programmes.

Chapter 10

Industry perspective and conclusions

Mark Shelbourn, Dino Bouchlaghem and Patricia Carrillo

Introduction

This concluding chapter presents further industry outlook on collaborative working based on the general perception of practitioners on the effects of the lack of proper planning and management of collaboration in projects. This was then followed by an evaluation exercise in which experienced construction industry personnel provided feedback on the benefits that the Planning and Implementation of Effective Collaboration in Construction (PIECC) framework, introduced and described in Chapter 3, can bring to current practice. Finally some overall conclusions drawn from the material presented in this book are presented.

Industry perspective on effective collaboration

A one-day workshop with industry representatives was organised to discuss practical issues related to collaborative working. The aim of the workshop was to capture and analyse the impact and implications of ineffective collaboration on project performance, and how the PIECC framework presented in Chapter 3 could help address these impacts. The third aspect concentrated on the effect of introducing the PIECC framework into construction projects. All participants in the workshop had extensive experience in collaborative working in the construction sector and represented a cross-section of the construction sector including architects, design managers, project managers, technology providers and some with a combined construction and technology experience.

The main activities of the workshop were conducted in groups and split into two distinct but interrelated tasks:

1 brainstorming and analysing the impact of ineffective collaborative working on project performance from the business, project or technology perspective;
2 evaluation of the potential benefits of using the PIECC framework to improve project performance.

For the first task, the groups were given an hour to identify the five most critical impacts of ineffective collaborative working on project performance. The group sessions were facilitated by industry representatives who collaborated in the PIECC project and the results were recorded by a member of the research team. In the second task, the groups were then asked to highlight areas of the PIECC framework that could be used to address these impacts to enable more effective collaborative working on projects. The results from the two tasks are described in the next section.

Task 1 results

Each of the three groups addressed the impact of ineffective collaborative working on project performance and focused on one of the following areas: business, project and technology. Table 10.1 shows a summary of the results from each group.

The table shows that there are commonalities in the issues identified in the business and project areas. For example, the most important issue highlighted was related to building trust in the project team; where this is found to be ineffective, there is a clear breakdown of relationships between collaborators in the team. Projects can also suffer from 'cost and time inefficiencies due to wasted effort and repetition of work', which in turn leads to the team being affected by a 'lack of focus' on the overall goals and main outputs of the venture. Associated with this, many projects experience 'too much waste', which leads to a 'loss of value and quality' in projects. Other issues identified include a lack of 'engagement' and 'ownership' within project teams, leading to poor morale and low motivation. This can then lead to a lack of knowledge sharing and innovation, resulting in limited creativity by project teams.

The main technological issues that result from ineffective collaborative working in projects included 'poor communication' as a result of 'lack of standards' and 'collaborative cultures and contracts'. The barriers associated with the technology issues were seen by the workshop participants to be much easier to break down than the project and business aspects.

Table 10.1 Summary of the results from task 1 of the workshop

Business	*Project*	*Technology*
Poor morale and motivation	Breakdown of relationships	Poor communication
Lack of focus	Cost + time inefficiency due to waste of effort/repetition	Lack of collaborative contracts and cultures
Too much waste	Loss of value/quality	Lack of standards
Barriers to creativity	Lack of ownership/poor morale/loss of engagement	Working in 3D–*n*D
Managing risk and creating trust	Lack of knowledge sharing and innovation	Differences in software functions of vendors

Tables 10.2, 10.3 and 10.4 present a detailed breakdown of the impacts identified in task 1.

After further analysis and synthesis of the data collected from task 1 of the workshop there was a surprising similarity between the results obtained from the business and project groups, but not so much overlap between these two groups and the technology group. A mapping of the connections between the issues identified by the three groups is shown in Figure 10.1.

Table 10.2 Top five effects on collaborative working from the business perspective

Poor morale and motivation	A resistance from people to change A need to be more proactive rather than reactive A need to plan early to get everybody on board Needs executive management Make the collaboration an enjoyable process A lack of rewards for collaborative working What are the sources of motivation for people, projects or organisations – the differences = difficulties for effective collaborative working
Lack of focus	Keeping people engaged who are off site/in peripheral areas Cannot say we can deliver X for Y cost because there is no way of focusing on the business drivers On own issues – not the projects Project extranets are a barrier for collaborative working Poor communication of processes to all levels of the organisation Ineffective collaboration takes away the focus from the overall end product and client satisfaction – 'we take our eye off the ball'
Too much waste	People work inefficiently Duplication of effort when acting as individual organisations The industry is fragmented in nature – this works against getting a consistent group of people to work on projects The industry is motivated by flexibility – do not want 'all the eggs in one basket' – business imperative to work with everyone 'Inertia of efficiency' – each company has own view of efficient working – hard to get them to give it up for collaboration Focus on individual commissions, individual projects, individual . . .
Barriers to creativity	Making collaborative working an enjoyable experience Seeking the optimum solution Restriction on 'adding value' Not wanting to risk it all 'Blame culture' – people are always preparing reasons for failure Difficulty of selling benefits to individual contractors – especially the value of investing effort
Managing risk/trust	A resistance from people to change A lack of understanding of others' 'drivers, incentives, motivation' Effect on the 'bottom line' – the return is not there if collaborative working is not effective Extra time at the front end will avoid problems further down the line 'Blame culture' – people are always preparing reasons for failure Lack of a sense of shared responsibility for the project outcomes

Table 10.3 Top five effects on collaborative working from the project perspective

Breakdown of relationships	Mismatched professional values: a lack of understanding each other, creates suspicion, prejudices surface Roles and responsibilities: lack of clarity Lack of early involvement: poor communication Coordination of process: conflicts between team members Coordination of information: relationship breakdown Poor communication: affects the morale of the team and client relationships
Cost + time inefficiency due to waste of effort/ repetition	Roles and responsibilities: mismatches lead to repetition of effort and wasted communication Lack of early involvement: means repetition of effort from the legacy of earlier decisions, increased risk of mistakes and rework Coordination of information: reinventing/rework means programme slippage – costs and sequence, errors, affecting health and safety Poor communication: increases costs and misunderstanding leading to wasted effort/mistakes/errors/omissions/duplication Loss of client value: costs to the client
Loss of value/quality	Mismatched professional values: different priorities Roles and responsibilities: lack of clarity Lack of early involvement: difficult to recognise value Coordination of process: quality suffers Coordination of information: quality of output Loss of client value: decreased customer satisfaction, meaning a poor reputation, lower share price and image
Lack of ownership/ poor morale/loss of engagement	Mismatched professional values: a lack of enjoyment, people's inhibitions increase with a reluctance to accept change Roles and responsibilities: it is always somebody else's problem, poor morale Lack of early involvement: lack of ownership and influence is diminished Poor communication: morale of the team diminishes leading to misaligned expectations
Lack of knowledge sharing and innovation	Roles and responsibilities: wasted communication, leading to a loss of momentum Lack of early involvement: stifles innovation/value engineering and management Coordination of process: information retrieval Loss of client value: no repeat business

Having determined the negative impacts that ineffective collaborative working can have on project performance in construction, the workshop then progressed to the second task of the day. Here the groups used the PIECC framework to determine which of the 23 processes could be used to address the issues identified in task 1. The findings from this task are summarised in the next section.

Table 10.4 Top five effects on collaborative working from the technology perspective

Poor communication	Using and reusing project information
Lack of collaborative contracts and cultures	No industry standard ones for teams to use A need to move towards single-project insurances and bank accounts
Lack of standards	A need to be more prescriptive Redefining the BS 1192 Part 5 Differences in the interpretation of the available standards
Working in 3D–*n*D	Most of the industry is still working in 2D, electronic paper and electronic filing cabinets Lack of 'template' files available – they do not come with the technology
Differences in software functions of vendors	A lack of understanding from the users Users are not properly educated/trained to use the available technology

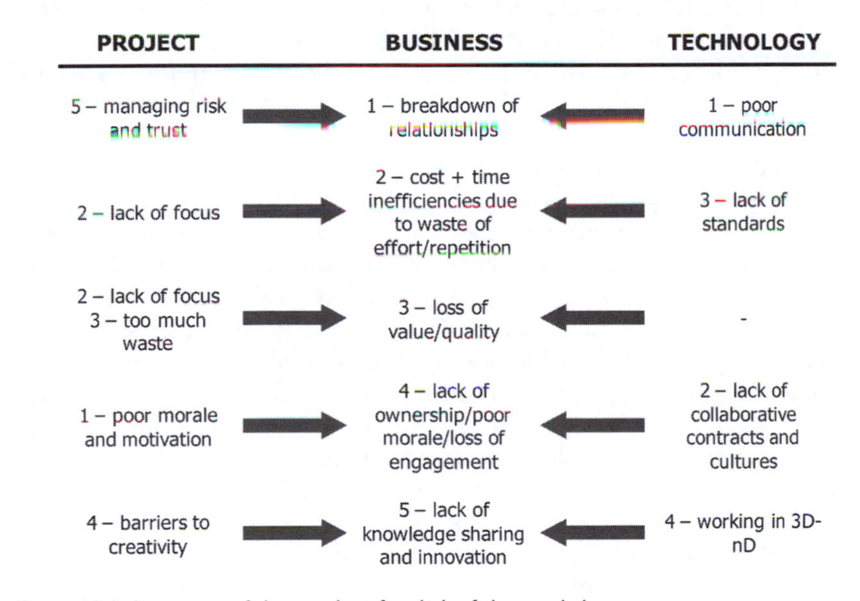

Figure 10.1 Summary of the results of task 1 of the workshop.

Task 2 results

The second task of the day was to 'assess the potential benefits of using the PIECC framework' to address the inefficiencies in collaborative working identified in task 1. The participants remained in the same groups and again were given one hour to conduct the task. A rapporteur from each group then presented the findings to the participants of the workshop. Collation of the information gathered shows the areas where the PIECC framework could break down the inefficiencies in current collaborative working practices in the following areas (the following section refers back to the PIECC framework presented in Chapter 3):

1 *The business perspective*
From a business perspective, the 'develop a shared vision' and 'measure collaboration performance' activities can be used to provide a reward and recognition policy to address the poor motivation and morale aspects. The lack of focus can be addressed by the development of a shared business strategy in the early stages of the collaborative working venture (see Figure 3.3). Other aspects of the framework that promote a shared focus include the development of a shared vision and assessing the risks of the venture. Barriers to creativity in the venture can be dealt with by carefully managing the fluidity of the project team; bringing the right people together in the first instance; and defining their roles and responsibilities at the outset of the venture. Also closely aligned to this aspect is how the PIECC framework can manage collaborative working risks and building trust within the team. It does this by imparting information on dispute resolution and providing managers with tools to build shared responsibility into the processes and procedures developed for their collaborative working project.
2 *The project perspective*
The results from a project perspective generally follow a similar pattern to the business section. To combat the breakdown of relationships, the PIECC framework suggests refining the business-level strategy, team selection and allocating appropriate resources in the business strategy section. In the collaboration brief section, start-up workshops can help to define the ground rules, project charter and success criteria and establish common goals. In planning the solution, individuals' roles and responsibilities and transparency of performance can be established. Implementing the solution refines the business-level strategy, team selection and resource allocation to reflect the developments in the collaborative working strategy for the project.
3 *Cost and time inefficiencies*
Cost and time inefficiencies and wasted effort are reduced by identifying partners that can work well together, providing them with the tools to set off in the same direction together by focusing on processes, procedures and core competencies that enable the adoption of more efficient ways of working. Clarity of purpose is developed to enable efficient collaborative working within the team.
4 *Value and quality*
The potential for loss of value and quality is addressed by developing a shared business strategy that is focused on output criteria identified at

the outset of the project. The PIECC framework enables team goals to be tied to the client values through the development of a collaboration brief. This means that the client's values are built into the processes and procedures to be adopted in the collaborative working venture. It is important to stress the importance of these values in the briefings and inductions of project team members.

5 *Lack of ownership*
The lack of ownership, poor morale and loss of engagement can be addressed early on in the development using the PIECC framework. The bringing together of the business strategies of all organisations involved in the collaborative working venture and deciding upon a 'champion' to lead the collaboration is perhaps the most important step in the process. Other aspects that need to be addressed include establishing the right culture for the project and, in developing the shared vision, bearing in mind that it should be positive and should set the behavioural rules (modus operandi) of the project. Ownership is built into the roles and responsibilities and personal development plans of the people involved in the venture; these should also be included in the codes of conduct for the collaboration.

6 *Lack of knowledge sharing and innovation*
The lack of knowledge sharing and innovation aspects are addressed by selecting organisations to work with that have a proven track record in knowledge sharing and innovation. The collaboration strategy allows the development of roles and targets that define agreements for incentive schemes, and knowledge capture and sharing processes. These roles and targets should be communicated to the team when they are brought together as part of training sessions for the collaborative working venture.

The technological aspects are addressed by the PIECC framework in less detail than the people and business aspects. The use of the PIECC framework would enable 'poor communications' to be removed. The development of a 'collaborative culture' is achieved by sharing goals, objectives and resources between the participating organisations. The other issues related to the development of standards, the software functionalities of the information and communication technology (ICT) providers, and whether or not to work in 3D–*n*D are all outside the scope of the PIECC framework. The PIECC work acknowledges initiatives that are addressing these ICT aspects and are included in the detailed information that supports the framework.

Summary of the evaluation

Figure 10.2 ranks the PIECC framework processes in order of importance in enabling a successful planning and implementation of effective

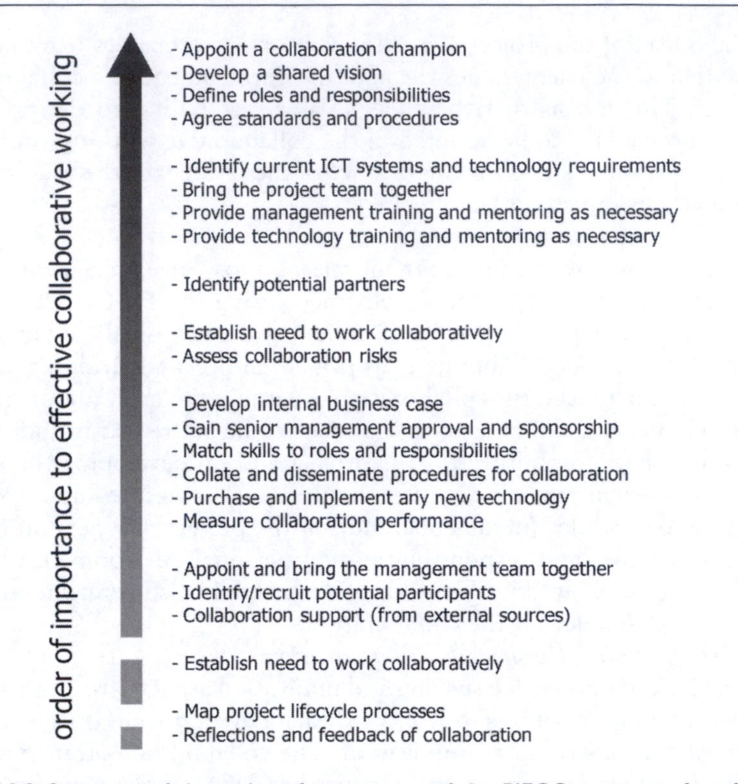

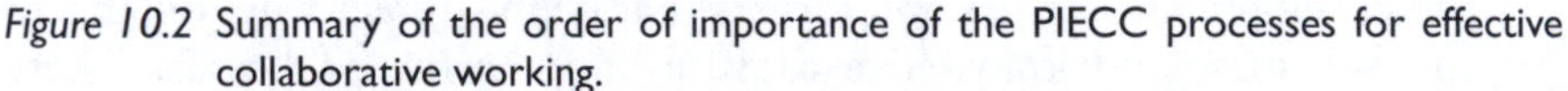
Figure 10.2 Summary of the order of importance of the PIECC processes for effective collaborative working.

collaborative working in projects. According to the evaluation results, the most important processes in the PIECC framework are 'appoint a collaboration champion', 'develop a shared vision', 'define roles and responsibilities' and 'agree standards and procedures'. However, it should be stressed that all the processes of the PIECC framework are to be considered together for the success of collaborative working.

The concluding discussions during the industrial workshop indicated that the PIECC project could improve team performance in collaborative working. Using Bruce Tuckman's model for building successful teams shown in Figure 10.3 (Smith, 2005), it was felt that if the PIECC framework were used it would probably reduce the distance between the teams' 'enthusiasm' and 'performance' during the 'forming' and 'storming' phases of the team development. Therefore Tuckman's model would look significantly different as a result (Tuckman, 1965; see Figure 10.4).

Perceived benefits of the PIECC framework

From the evaluation and validation exercise the list below summarises some of the perceived benefits of the PIECC framework:

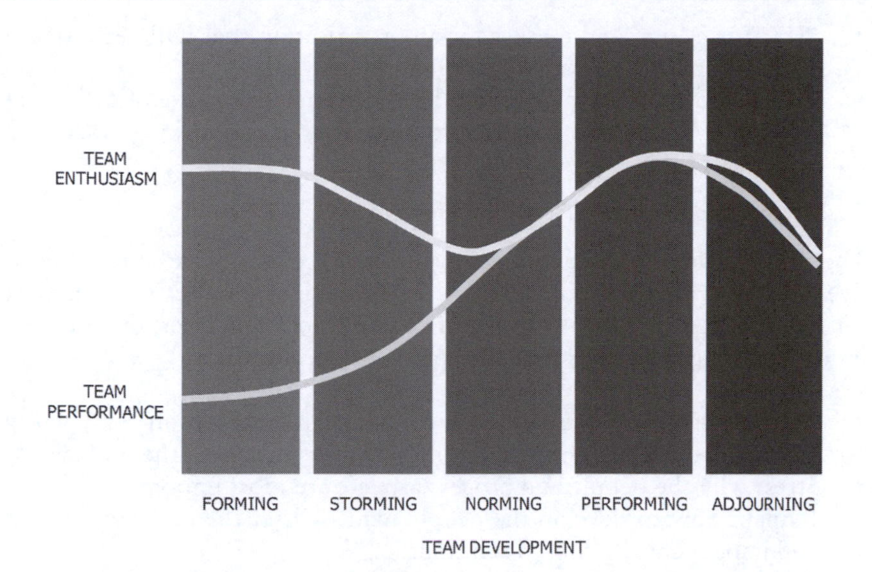

Figure 10.3 Tuckman's model of team development and its effect on team enthusiasm and performance (Tuckman, 1965).

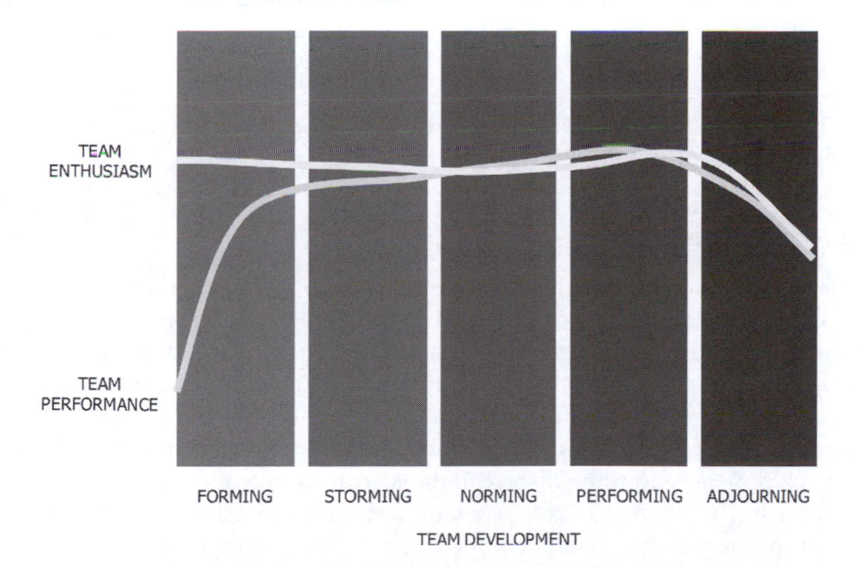

Figure 10.4 Tuckman's model of team development and its effect on team enthusiasm and performance if the PIECC methodology were used.

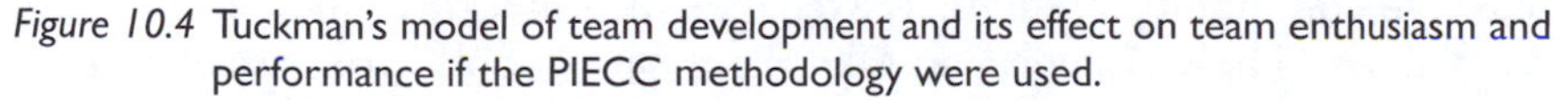

- PIECC is a 'good framework to provide people with the tools to develop a strategy for collaboration'.
- PIECC 'really makes you think about collaboration'.
- PIECC highlights 'the interrelationships needed to ensure that a route is followed to plan and implement collaborative working'.

- PIECC is a 'sign and marker for an organisation that will need to win work and carry out that work collaboratively'.
- PIECC prevents 'people [giving] lip service to the uptake of collaboration, only for them to [go] back to normal/original working practices'.

Further work and recommendations for improvements

The results from the evaluation and validation exercise have shown a positive response from the industry regarding the value of the PIECC framework for effective collaborative working in construction. However, there were a number of suggested improvements and considerations:

- Add references to the Belbin and European Foundation for Quality Management models into the additional information for the framework.
- Stress why the people and process aspects are most important.
- Explain how to develop the people who will set the collaboration up and implement the chosen solution.
- Evaluate and validate the framework in real construction projects/case studies.
- Should PIECC be project-based or organisation-based?
- Include a method for measuring the effectiveness of collaborative working.

These suggestions and considerations will be taken up in any future developments of the PIECC framework.

Overall conclusions

The work presented in this book was prompted by the current limitations and shortfalls of the approaches used for collaborative working in the construction industry. The fragmented nature of construction, in which each project is a one-off in terms of objectives, final product and partnerships, creates a business environment that is not conducive to long-term collaborative arrangements. The organisations that come together for the delivery of each project tend to be diverse and disparate and enter into a temporary partnership to work towards a set of common objectives defined by the client. Managing this partnership requires systems that enable distributed teams to communicate, interact, exchange information and coordinate the work flows. These systems can take the form of tools, processes and procedures and need to be planned and implemented in an integrated manner to ensure a successful collaboration and deliver a product that meets the initial project goals. However, and despite the availability of sophisticated collaboration technologies, the reality of construction projects shows that effective well-planned collaboration is still not fully realised because of the following:

- The benefits of properly planned collaboration are not yet being acknowledged on a wide scale. Contractors do not have established supply chains and suppliers are still contributing to projects on a one-off basis.
- Technological needs are still considered ahead of, and in isolation from, the requirements of the multidisciplinary business environment and the end users.
- Technologies for distributed collaboration and group decision support are used in an uncoordinated manner and have not been adopted widely. Although a variety of factors from cost to management support may be responsible for this, the most common factor is the lack of awareness by users of what these tools are, what they can do, and how or where they can best be applied. This reflects not only the lack of involvement from end users but also the absence of an integrated requirement capture process that takes into account the business and process factors.
- Most companies still do not have well-defined methods for selecting their collaborators and solely rely on competitive tendering. As a result of this, they neither have full knowledge of whom they enter into partnership with nor understand their underlying processes. The selection of collaboration partners needs to be based on both capability and cost competitiveness.
- Organisations in the construction industry lack consistent and well-planned work streams. Collaborative arrangements can provide continuity of work for selected partners within defined framework agreements. From an individual partner perspective, it is advantageous to have a pool of known collaborators to work with on different projects and therefore implement better-integrated design and construction solutions.
- Experience shows that collaborators usually indicate their willingness to work a certain way even though they have little experience of doing so. This can be avoided through carefully planned collaborative arrangements whereby partners are able to improve their performance and profits from project to project.

In response to the above shortfalls, this book presented a number of frameworks and methods to support various aspects of collaboration in the construction industry. These include:

- a strategic decision-making methodology that guides organisations in the planning for effective collaborative working and the implementation of suitable systems, tools and techniques (Chapter 3);
- a structured and practical approach for the implementation of a holistic information management strategy to enable collaborative working within an organisation or within a specific construction project (Chapter 4);

- a set of practical scenarios for collaboration environments that integrate mobile and wireless technologies to support and facilitate collaboration between designers in design offices and construction teams on site (Chapter 5);
- a multidisciplinary collaborative method that integrates design objectives and coordinates the trade-offs between the design disciplines in the process of identifying an optimum design solution (Chapter 6);
- a framework for the management of change when implementing collaboration environments (Chapter 8).

In addition, the theories and concepts related to collaborative working and change management are reviewed in Chapters 2 and 7. Finally industry perspective and case studies on the planning and implementation of collaborative working in practice were presented in Chapters 9 and 10.

References

Alexandrov, N.M., Lewis, R.M. (2002) Analytical and computational aspects of collaborative optimization and multidisciplinary design. *AIAA Journal*, 40: 301–309.

Al-Homoud, M.S. (2005) A systematic approach for the thermal design optimisation of building envelopes. *Journal of Building Physics*, 29: 95–119.

Al-Mashari, M., Irani, Z., Zairi, M. (2001) Business process reengineering: a survey of international experience. *Business Process Management*, 7(5): 437–455.

Alvarez, R. (2001) 'It was a great system': face work and discursive construction of technology during information systems development. *Information, Technology & People*, 14(4): 385–405.

Andersson, C. (2001) *GPRS and 3G Wireless Applications*, John Wiley & Sons: New York.

Anumba, C.J., Baron, G., Evbuomwan, N.F.O. (1997) Communications issues in concurrent life-cycle design and construction. *BT Technology Journal*, 15(1): 209–216.

Anumba, C.J., Bouchlaghem, N.M., Whyte, J. (2000) Perspectives on an integrated construction project model. *International Journal of Cooperative Information Systems*, 9(9): 283–313.

Anumba, C.J., Ugwa, O.O., Newnham, L., Thorpe, A. (2002) Collaborative design of structures using intelligent agents. *Automation in Construction*, 11(1): 89–103.

Anumba, C.J., Ren, Z., Thorpe, A., Ugwu, O.O., Newnham, L. (2003) Negotiation within a multi-agent system for the collaboration design of light industrial buildings. *Advance in Engineering Software*, 34: 389–401.

Atkin, B., Borgbrant, J., Josephson, P.E. (2003) *Construction Process Improvement*, Blackwell Publishing: Oxford.

Argyris, C. (1985) *Strategy, Change and Defensive Routines*, Pitman: Boston.

Attaran, M. (2000) Why does reengineering fail? A practical guide for successful implementation. *Journal of Management Development*, 19(9): 794–801.

Balling, R.J, Rawlings, M.R. (2000) Collaborative optimisation with disciplinary conceptual design. *Structural Multidisciplinary Optimisation*, 20: 232–241.

Balogun, J., Hope Hailey, V. (2004) *Exploring Strategic Change*, 2nd edition, FT Prentice Hall: Hemel Hempstead.

Bartoli, A., Hermel, P. (2004) Managing change and innovation in IT implementation process. *Journal of Manufacturing Technology Management*, 15(5): 416–425.

Bentley, J.E. (2001) *Metadata: Everyone Talks About It, but What Is It?*, http://www.sas.com.

Ben-Yitzhak, O., Golbandi, N., Har-El, N., Lempel, R. (2008) Beyond basic faceted search. *Proceedings of the International Conference on Web Search and Web Data Mining*, 11–12 February, Palo Alto, CA.

Bjork, B., Huovila, P., Hult, S. (1993) Integrated construction project document management. *Proceedings of the EuropIA'93 Conference*, Delft, Netherlands.

Bouchlaghem, D. (2002) Optimising the design of building envelopes for thermal performance. *Automation in Construction*, 10(1): 101–112.

Bouchlaghem, D., Shang, H., Whyte, J., Ganah, A. (2005) Visualisation in architecture, engineering and construction. *Automation in Construction*, 14(3): 287–295.

Bouchlaghem, D., Shang, H., Anumba, C.J., Miles, J., Cen, M., Taylor, M. (2006) An ICT enabled collaborative working system for concurrent conceptual design. *International Journal of Architectural Engineering and Design Management*, 1(4): 261–280.

Bowden, S., Dorr, A., Thorpe, T., Anumba, C. (2006) Mobile ICT support for construction process improvement. *Automation in Construction*, 15(5): 664–676.

Braun, R.D., Kroo, I.M. (1995) Development and application of the collaborative optimisation architecture in multidisciplinary design environment. Technical report, NASA Langley Technical Report Server.

Bresnen, N., Marshall, N. (2000) Building partnerships: case studies of client–contractor collaboration in the UK construction industry. *Construction Management and Economics*, 18: 819–832.

Bridges, J.D. (2007) Taking ECM from concept to reality. *Information Management*, 41(6): 30–39.

Broughton, V., Slavic, A. (2007) Building a faceted classification for the humanities: principles and procedures. *Journal of Document*, 63: 727–754.

Bruce, W.R., Gilster, R. (2002) *Wireless LANs: End to End*, John Wiley & Sons: New York.

Brunner, S., Ali, A.A. (2004) Voice over IP 101. *Understanding VoIP Networks*, White Paper, Juniper Networks, http://www.juniper.net.

Buchanan, S., Gibb, F. (1998) The information audit: an integrated strategic approach. *International Journal of Information Management*, 18(1): 29–47.

Burnes, B. (1996) No such thing as . . . a 'one best way' to manage organizational change. *Management Decision*, 34(10): 11–18.

Burns, T., Stalker, G.M. (1961) *The Management of Innovation*, Tavistock: London.

Caldas, H.C., Soibelman, L. (2003) Automating hierarchical document classification for construction management information systems. *Journal of Automation in Construction*, 12: 395–406.

Cao, G., Clarke, S., Lehaney, B. (2000) A systemic view of organizational change and TQM. *TQM Magazine*, 12(3): 186–193.

Cao, G., Clarke, S., Lehaney, B. (2001) A critique of BPR from a holistic perspective. *Business Process Management Journal*, 7(4): 332–339.

Carnall, C.A. (1990) *Managing Change in Organizations*, Prentice Hall: Hemel Hempstead.

Chaffey, D., Wood, S. (2004) *Business Information Management: Improving Performance Using Information Systems*. Pearson Education: Harlow.

Chandler, A.D. (1962) *Strategy and Structure: Chapters in the History of the Industrial Enterprise*, MIT Press: London.

Chen, J., Zhong, Y., Xiao, R., Sun, J. (2005) The research of the decomposition-coordination method of multidisciplinary collaborative design optimisation. *International Journal for Computing-Aided Engineering and Software*, 22: 274–285.

Cheng, E.W.L., Li, H., Love, P.E.D., Irani, Z. (2001) An e-business model to support supply chain activities in construction. *Logistics Information Management*, 14(1/2): 68–77.

Chesterman, D. (2001) Learning from research perspectives in collaborative working. *Career Development International*, 6(7): 378–383.

Child, J. (1973) Parkinson's progress: accounting for the number of specialists in organizations. *Administrative Science Quarterly*, 18(3): 328–348.

Choudhary, R., Michalek, J. (2005) Design optimization in computer aided architectural design. *International Conference of Association for Computer Aided Architectural Design Research in Asia*, 28–30 April, New Delhi, India.

Choudhary, R., Malkawi, A., Papalambros, Y.P. (2005) Analytic target cascading in simulation-based building design. *Automation in Construction*, 14: 551–568.

Christian, M. (2002) Conquering business challenges with ECM: the agile and efficient corporation. *Best Practices in Enterprise Content Management*, 11(3).

CII (Construction Industry Institute) (1994) *Project Change Management*, Special Publication 43-1.

Claver, E., Llopis, J., Reyes Gonzalez, M., Gasco, J.L. (2001) The performance of information systems through organisational change. *Information Technology & People*, 14(3): 247–260.

COCONET (2003) *Research Agenda and Roadmap: Context-Aware Collaborative Environments for Next-Generation Business Networks (COCONET IST-2001–37460)*, http://www.europa.net.

Coffey, A., Atkinson, P. (1996) *Making Sense of Qualitative Data*, Sage: Thousand Oaks, CA.

Cohen, J. (2000) *Communication and Design with the Internet*, W. W. Norton & Co.: New York.

Compaq Official Web Site (2004) http://www.compaq.com.

Computer Associates International (2002) *Enabling Mobile eBusiness Success*, White Paper, http://www.ca.com.

Conner, D.R. (1993) *Managing at the Speed of Change*, Villard Books: New York.

Conner, M., Finnemore, P. (2003) Living in the new age: using collaborative digital technology to deliver health care improvement. *International Journal of Health Care Quality Assurance*, 16(2): 77–86.

Cox, I.D., Morris, J.P., Rogerson, J.H., Jared, G.E. (1999) A quantitative study of post contract award design change in construction. *Construction Management & Economics*, 17: 427–439.

Credé, A. (1997) Social, cultural, economic and legal barriers to the development of technology-based information systems. *Industrial Management & Data Systems*, 97(1): 58–62.

Crosby, P. (1994) *Completeness: Quality for the 21st Century*. Plume: New York.

Cummings, T.G., Worley, C.G. (2005) *Organization Development and Change*, 8th edition, South Western: Mason, OH.

Davenport, T. (2000) Putting the I in IT. In Marchand, D.A., Davenport, T.H., Dickson, T. (eds), *Mastering Information Management*, Pearson Education: Harlow.

Davenport, T., Marchand, D. (2000) Is knowledge management just good information management? In Marchand, D.A., Davenport, T.H., Dickson, T. (eds), *Mastering Information Management*, Pearson Education: Harlow.

Day, M. (2006) Metadata: a general introduction. *Cataloguing Online Resources Conference*, Manchester, UK.

Deb, K., Pratap, A., Agarwal, S., Meyarivan, T. (2002) A fast and elitist multi-objective genetic algorithm: NSGA-II. *IEEE Transactions on Evolutionary Computation*, 6: 182–197.

DeRoure, D., Hall, W., Reich, S., Pikrakis, A., Hill, G., Stairmand, M. (1998) An open framework for collaborative distributed information management. *Computer Networks and ISDN Systems*, 30: 624–615.

Dilson, A. (2002) Information architecture in JASIST: Just where did we come from? *Journal of the American society for Information Science and Technology*, 53(10): 821–823.

Donaldson, L. (2001) *The Contingency Theory of Organizations*. Sage: Thousand Oaks, CA.

Dong, A., Agogino, A. (2001) Design principles for the information architecture of a SMET education digital library. *First Joint Conference on Digital Libraries*, Roanoke, VA.

DTI (n.d.) *VoIP Factsheet*, http://www.dti.gov.uk/bestpractice.

Dustdar, S., Gall, H. (2003) Architectural concerns in distributed and mobile collaborative systems. *Journal of Systems Architecture*, 49: 457–473.

EEDO knowledgeware (2006) *Metadata, Taxonomies and Content Re-usability*, White Paper, http://www.adlcommunity.net.

Egan, J. (1998) *Rethinking Construction*, HMSO: London.

Elvin, G. (2003) Tablet and wearable computers for integrated design and construction. *Construction Research Congress 2003*, Honolulu, HI.

Emmitt, S., Gorse, C. (2003) *Construction Communication*, Blackwell Publishing: Oxford.

Eseryel, D., Ganesan, R., Edmonds, G. (2002) Review of computer-supported collaborative work systems. *Educational Technology and Society*, 5(2): 130–136.

Evgeniou, T., Cartwright, P. (2005) Barriers to information management. *European Management Journal*, 23(3): 293–299.

Faniran, O., Love, P.E.D., Treloar, G., Anumba, C.J. (2001) Methodological issues in design–construction integration. *Logistics Information Management*, 14(5/6): 421–426.

Felkins, P.K., Chakiris, B.J., Chakiris, K.N. (1993) *Change Management: A Model for Effective Organizational Performance*, Quality Resources: White Plains, NY.

Ferneley, E., Lima, C., Fies, B., Rezgui, Y., Wetherill, M. (2003) Inter-organisational semantic webs to enable knowledge discovery and dissemination: technical support for the social process. *Proceedings of the 10th ISPE International Conference on Concurrent Engineering (CE 2003)*, Madeira.

Finne, C. (2003) How the internet is changing the role of construction information middlemen: the case of construction information services. *ITcon*, 8: 397–411.

Fonseca, C.M., Fleming, P.J. (1993) Multi-objective genetic algorithm. *Proceedings of IEEE Colloquium on Genetic Algorithm for Control Systems Engineering*, 28 May, London.

Ford, J.D., Ford, L.W., McNamara, R.T. (2002) Resistance and the background conversations of change. *Journal of Organizational Change Management*, 15(2): 105–121.

Gibbs, G.R. (2002) *Qualitative Analysis: Explorations with NVivo*, McGraw-Hill: Maidenhead.

Giess, M.D, Wild, P.J, McMahon, C.A. (2008) The generation of faceted classification schemes for use in the organisation of engineering design documents. *International Journal of Information Management*, 28: 379–390.

Gottlieb, S. (2005) From enterprise content management to effective content management. *Cutter IT Journal*, 18(5): 13–18, http://www.cutter.com.

Graetz, F., Rimmer, M., Lawrence, A., Smith, A. (2006) *Managing Organizational Change*, 2nd edition, John Wiley & Sons: Milton, QLD.

Grosniklaus, M., Norrie, M.C. (2002) Information concepts for content management. *Proceedings of the Third International Conference on Web Information Systems Engineering*, Singapore.

GSM Association (n.d.) *GSM: The Wireless Evolution*, http://www.gsmworld.com/technology/index.shtml.

Gyampoh-Vidogah, R., Moreton, R. (2003) Implementing information management in construction: establishing problems, concepts and practice. *Construction Innovation*, 3: 157–173.

Hamer, C.E. (2006) Six Sigma reasons to embrace enterprise content management, from customer centric content management. *Rockley Report Newsletter*, Rockley group.

Haque, B., Pawar, K.S. (2003) Organisational analysis: a process-based model for concurrent engineering environments. *Business Process Management Journal*, 9(4): 490–526.

Hearst, M. (2006) Design recommendations for hierarchical faceted search interfaces. *Proceedings from ACM SIGIR Workshop on Faceted Search*, 10 August, Seattle, WA.

Hicks, B.J, Culley, S.J, Allen, R.D., Mullineux, G. (2002) A framework for the requirements of capturing storing and reusing information and knowledge in engineering design. *International Journal of Information Management*, 22: 263–280.

Hicks, B.J., Culley, S.J., McMahon, C.A. (2006) A study of issues relating to Information Management across engineering SMEs. *International Journal of Information Management*, 26: 267–289.

Hienrich, J., Pipek, V., Wulf, V. (2005) Context grabbing: assigning metadata in large document collections. *Proceedings of the Ninth European Conference on Computer-Supported Co-operative Work*, 18–22 September, Paris, France.

Hewlett Packard (2007) *Managing Data as a Corporate Asset: Three Action Steps towards Successful Data Governance*, White Paper, http://www.hp.com.

Hsieh, T., Lu, S., Wu, C. (2004) Statistical analysis of causes for change orders in metropolitan public works. *International Journal of Project Management*, 22(8): 679–686.

Humphrey, W.S. (1987) *Characterizing the Software Process: A Maturity Framework*, Software Engineering Institute, Carnegie Mellon University: Pittsburgh.

Humphrey, W.S. (1988) Characterizing the software process. *IEEE Software*, 5(2): 73–79.

Ibbs, C.W., Wong, C.K., Kwak, Y.H. (2001) Project change management system. *Journal of Management in Engineering*, 17(3): 159–165.

iDesk (n.d.) *VoIP Brochure*, http://www.idesk.com/downloads/Voip%20Brochure.pdf.

Intel (2004) *Broadband Wireless: The New Era in Communications*, White Paper, http://www.intel.com.

Intel (2003) *Deploying Secure Wireless Networks*, White Paper, http://www.intel.com.

Intel (2005) *Transforming Communications through End-to-End Voice Over IP Solutions*, http://www.intel.com.

Intel and WiMAX (2003) *IEEE 802.16 and WIMAX, Broadband Wireless Access for Everyone*, White Paper, http://www.intel.com.

Intel and WiMAX (2004) *Understanding WIMAX and 3G for Portable/Mobile Broadband Wireless*, Technical White Paper, http://www.intel.com.

Intel and WiMAX (2005) *Accelerating Wireless Broadband*, http://www.intel.com.

International Standards Office (2001) *ISO 15489-1: Information and Documentation – Records management. General*, ISO: Geneva.

Irani, Z., Beskese, A., Love, P.E.D. (2004) Total quality management and corporate culture: constructs of organisational excellence. *Technovation*, 24(8): 643–650.

Jick, T.D., Peiperl, M.A. (2003) *Managing Change: Cases and Concepts*, McGraw Hill: Boston.

Jimenez, F., Gomez-Skarmeta, A.F., Sanchez, G., Deb, K. (2002) An evolutionary algorithm for constrained multi-objective optimisation. *Proceedings of 2002 World Congress on Computational Intelligence*, 12–17 May, Honolulu, HI.

Jorgensen, B., Emmitt, S. (2009) Investigating the integration of design and construction from a 'lean' perspective. *Construction Innovation*, 9: 225–240.

Kalay, Y.E. (1999) The future of CAAD: from computer-aided design to computer-aided collaboration. *CAAD Futures 1999*, Atlanta, GA.

Kalay, Y.E. (2001) Enhanced multi-disciplinary collaboration through semantically rich representation. *Automation in Construction*, 10(6): 741–755.

Karjalainen, A., Paivarinta, T., Tyrvainen, P., Rajala, J. (2000) Genre based metadata for enterprise document management. *Proceedings of the International Conference on System Sciences*, Hawaii.

Karlson, B., Bria, A., Lonnqvist, P., Norlin, C., Lind, J. (2003) *Wireless Foresight*, John Wiley & Sons: Chichester.

Karsten, H. (n.d.) Collaboration and collaborative information technology: what is the nature of their relationship? Unpublished paper, University of Jyväskylä, Finland.

Kast, F.E., Rosenzweig, J.E. (1974) *Organization and Management: A Systems Approach*, McGraw-Hill: New York.

Khun-Jush, J., Malmgren G., Schramm, P., Torsner, J. (2000) HiperLAN Type 2 for broadband wireless communication. *Ericsson Review*, 2: 108–119.

Kitchen, P.J., Daly, F. (2002) Internal communication during change management. *Corporate Communications*, 7(1): 46–53.

Korac-Kakabadse, N., Kouzmin, A. (1999) Designing for cultural diversity in an IT and globalizing milieu: some real leadership dilemmas for the new millennium. *Journal of Management Development*, 18(3): 291–319.

Koschmann, T., Kelson, A.C., Feltovich, P.J., Barrows, H.S. (1996) Computer-supported problem based learning: a principled approach to the use of computers in collaborative learning. In T. Koschmann (ed.), *CSCL: Theory and Practice*, Lawrence Erlbaum Associates: Mahwah, NJ.

Koseoglu, O., Bouchlaghem, D. (2008) Mobile visualisation for on-site collaboration. *International Journal of Interactive Mobile Technologies*, 2(4): 43–53.

Kotter, J.P., Cohen, D.S. (2002). *The Heart of Change: Real Life Stories of How People Change Their Organizations*, Harvard Business School Press: Boston.

Kumaraswamy, M.M., Chan D.W.M. (1998) Contributors to construction delays. *Construction Management and Economics*, 16: 17–29.

Kvan, T. (2000) Collaborative design: what is it? *Automation in Construction*, 9: 409–415.

Land Mobile (2006) HSDPA and the pricing conundrum. *Wireless Communications for Business*, October: 25.

Lang, S.Y.T., Dickinson, J., Buchal, R.O. (2002) Cognitive factors in distributed design. *Computers in Industry*, 48(1): 89–98.

Latham, M. (1994) *Constructing the Team*, HMSO: London.

Lau, H.Y.K., Mak, K.L., Lu, M.T.H. (2003) A virtual design platform for interactive product design and visualization. *Journal of Materials Processing Technology*, 139(1–3): 402–407.

Laudon, K., Laudon, J. (2009) *Management Information Systems*, 11th edition, Pearson Education: New York.

Lawrence, P.R., Lorsch, J.W. (1967) *Organization and Environment: Managing Differentiation and Integration*, Harvard University Press: Boston.

Lazarus, D., Clifton, R. (2001) *Managing Project Change: A Best Practice Guide*, Hobbs: London.

Lee, I. (2004) Evaluating business process-integrated information technology investment. *Business Process Management Journal*, 10(2): 214–233.

Lin, J.G. (2004) Analysis and enhancement of collaborative optimisation for multidisciplinary design. *AIAA Journal*, 42: 348–360.

Loosemore, M. (1998) Organisational behaviour during a construction crisis. *International Journal of Project Management*, 16(2): 115–121.

Love, P.E.D., Holt, G.D., Shen, L.Y., Li, H., Irani, Z. (2002) Using systems dynamics to better understand change and rework in construction project management systems. *International Journal of Project Management*, 20(8), 425–436.

McAllister, C.D., Simpson, T.W., Hacker, K., Lewis, K., Messac, A. (2005) Integrating linear physical programming within collaborative optimization for multi-objective multidisciplinary design optimization. *Structural Multidisciplinary Optimization*, 29: 178–189.

McNay, H.E. (2002) Enterprise content management: an overview. *Proceedings of the International Professional Communication Conference IPCC*, Portland, OR.

Manthou, V., Vlachopoulou, M., Folinas, D. (2004) Virtual e-Chain (VeC) model for supply chain collaboration. *International Journal of Production Economics*, 87: 241–250.

Marchand, D. (2000) *Competing with Information: A Manager's Guide to Creating Business Value with Information Content*, John Wiley & Sons: New York.

Mattessich, P.W., Monsey, B.R. (1992) *Collaboration: What Makes It Work?* Amherst H. Wilder Foundation: St. Paul, MN.

May, A., Carter, C. (2001) A case study of virtual team working in the European automotive industry. *International Journal of Industrial Ergonomics*, 27: 171–186.

Mendez, M.M. (2005) Wimax and mobile operators. *Ovum*, http://www.ovum.com,.
MIT DSM Research Group (2005) *MIT DSM Web Site*, http://www.dsmweb.org/.
Moore, C., Markham, R. (2002) *Enterprise Content Management: A Comprehensive Approach for Managing Unstructured Content*, Giga information management group White Paper, http://www.forrester.com.
Munkvold, B.E., Paivarinta, T., Hodne, K.A., Stangeland, E. (2003) Contemporary issues of enterprise content management: the case of Stat Oil. *Proceedings of the 11TH European Conference of Information Systems.*
Munkvold, B.E., Paivarinta, T., Hodne, K.A., Stangeland, E. (2006) Contemporary issues of enterprise content management. *Scandinavian Journal of Information Systems*, 18(2): 69–100.
Murphy, L.D. (2001) Addressing the metadata gap: ad-hoc digital documents in organisations. In Chin, A.G. (ed.), *Text Databases & Document Management: Theory and Practice*, Idea Group Publishing: Hershey, PA.
National Audit Office (2009) *The Building Schools for the Future Programme: Renewing the Secondary School Estate*, The Stationery Office: London.
Netlink Wireless Telephone Portfolio (n.d.) *Wi-Fi Telephony for the Enterprise*, http://www.spectralink.com.
NISO (National Information Standards Organisation) (2004) *Understanding Metadata*, NISO Press: Bethesda, MD.
NSF (National Science Foundation) (1996) Research opportunities in engineering. In Shah, J. (ed.), *NSF Strategic Planning Workshop Final Report*, http://www.nsf.gov.
O'Conner, P. (1994) Quality, reliability, and reengineering. *Quality and Reliability Engineering International*, 10(6): 451–452.
O'Connor, C.A. (1993) *The Handbook for Organizational Change Strategy and Skill for Trainers and Developers*, McGraw-Hill: Maidenhead.
Olofsson, T., Emborg, M. (2004) Feasibility study of field force automation in the Swedish construction sector. *Electronic Journal of Information Technology in Construction (ITcon)*, 9: 285–295.
Paganelli, F., Petenati, M.C., Giuli, D. (2006) A metadata based approach for unstructured document management in organisations. *Information Resource Journal*, 19(1): 1–22.
Paivarinta, T., Munkvold, B.E. (2005) Enterprise content management: an integrated perspective on Information Management. *Proceedings of the 38th International Conference on System Sciences*, Hawaii.
Paivarinta, T., Tyrvainen, P., Ylimaki, T. (2002) Defining organisational document metadata: a case beyond standards. *Proceedings from the European Conference on Information Systems, ECIS*, 6–8 June, Poland.
Papalambros, P.Y. (2002) The optimisation paradigm in engineering design: promise and challenges. *Computer-Aided Design*, 34: 939–951.
Patel, N.V., Irani, Z. (1999) Evaluating information technology in dynamic environments: a focus on tailorable information systems. *Logistics Information Management*, 12(1/2): 32–39.
Paton, R.A., McCalman, J. (2000) *Change Management: A Guide to Effective Implementation*, Sage: London.
Paulk, M.C., Curtis, B., Chrissis, M.B., Weber, C.V. (1993) *Capability Maturity Model for Software, version 1.1*, SEI, Carnegie Mellon University: Pittsburgh.
PC Magazine (2001) 3G: the next wave. March: 38–42.

Peansupap, V., Walker, D.H.T (2005) Factors enabling information and communication technology diffusion and actual implementation in construction organisations. *Electronic Journal of Information Technology in Construction*, 10: 193–218.

Pektas, S.T., Pultar, M. (2006) Modelling detailed information flows in building design with the parameter based design structure matrix. *Design Studies*, 27: 99–122.

Peña-Mora, F., Hussein, K., Vadhavkar, S., Benjamin, K. (2000) CAIRO: a concurrent engineering meeting environment for virtual design teams. *Artificial Intelligence in Engineering*, 14: 203–219.

Poulbere, V. (2005) Voice over WiFI: VOIP's first steps into the wireless area. *Ovum*, http://www.ovum.com.

Proctor, S., Brown, A.D. (1997) Computer-integrated operations: the introduction of a hospital information support system. *International Journal of Operations & Production Management*, 17(8): 746–756.

Proctor, T., Doukakis, I., (2003) Change management: the role of internal communication and employee development. *Corporate Communications: An International Journal*, 8(4): 268–277.

Proxim Wireless Networks (2004) *Wi-Fi in the Enterprise*, Position Paper, http://www.proxim.com.

Pushkar, S., Becker, R., Katz A. (2005) A methodology for design of environmental optimal buildings by variable grouping. *Building and Environment*, 40: 1126–1139.

Quirke, B. (1996) *Communicating Change*, McGraw-Hill: Maidenhead.

Ranganathan, S.R. (1963) *The Colon Classification*, 6th edition, Asia Publishing House: Kolkata.

Rebolj, D., Magdic, A., Babic, N. (2000) Mobile computing: the missing link to effective construction IT. Unpublished paper, Faculty of Civil Engineering, University of Maribor, Slovenia, http://www.fg.uni-mb.si/cgi/e-gradbisce/publikacije/Mobile%20computing%20-%20the%20missing%20link%20to%20effective%20construction%20IT.pdf.

Reimer, J. (2002) Enterprise content management. *Datenbank-Spektrum*, 2(4), 17–22.

Ringland, G. (2002) *Scenarios in Business*, John Wiley & Sons: Chichester.

Robey, D., Sales, C.A. (1994) *Designing Organizations*, Boston, EUA: Irwin, PA.

Robson, W. (1994) *Strategic Management and Information Systems: An Integrated Approach: The Management Information System Strategic Planning 'Toolkit'*, Pitman: Harlow.

Rockley, A. (2006) Customer centric content management. *Rockley Report*, 3(1).

Rockley, A., Kostur, P., Manning, S. (2003) *Managing Enterprise Content: A Unified Content Strategy*, 1st edition, New Riders: Indianapolis.

Roethlisberger, F.J., Dickson, W.J. (1939) *Management and the Worker*, Harvard University Press: Cambridge, MA.

Ross, J. (2003) Creating a strategic IT architecture competency: learning in stages, *MIS Quarterly Executive*, 2(1): 31–43.

Rouibah, K., Caskey, K.R. (2003) Change management in concurrent engineering from a parameter perspective. *Computers in Industry*, 50(4): 15–34.

Ruessmann, T., Preece, I., Peppard, J. (1994) Tools and methods in business process redesign. Working paper, Information Systems Research Centre, Cranfield School of Management.

Rye, C. (1996) *Change Management Action Kit*, Kogan Page: London.

Saidi, K., Haas, C.T., Balli, N.A. (2002) *The Value of Handheld Computers in Construction*, http://fire.nist.gov/bfrlpubs/build02/PDF/b02138.pdf.

Schaeffer, B. (2002) Navigating the content management jungle: a survival guide, *Intranet Journal*, http://www.intranetjournal.com.

Schwartz, P. (1999) *The Art of the Long View: Planning for the Future in an Uncertain World*, John Wiley & Sons: Chichester.

Senge, P.M. (1992) *The Fifth Discipline: The Art and Practice of the Learning Organization*, Century Business: London.

Senior, B., Fleming, J. (2006) *Organizational Change*, 3rd edition, Prentice Hall: Harlow.

Singh, P.J., Smith, A.J.R. (2004) Relationship between TQM and innovation: an empirical study. *Journal of Manufacturing Technology Management*, 15(5): 394–401.

Smith, M.E. (2003) Changing an organisation's culture: correlates of success and failure. *Leadership and Organisation Development Journal*, 24(5): 249–261.

Smith, M.K. (2005) Bruce W. Tuckman: forming, storming, norming and performing in groups. *The Encyclopaedia of Informal Education*, http://www.infed.org/thinkers/tuckman.htm.

Sprehe, J.T. (2005) The positive benefits of electronic records management in the context of enterprise content management. *Government Information Quarterly*, 22(2): 297–303.

Sterman, J.D. (2000) *Business Dynamics: Systems Thinking and Modeling for a Complex World*, McGraw-Hill: Boston.

Steward, W., Mann, B., Gilster, R. (2002) *Wireless Devices End to End*, Hungry Minds: New York.

Stouffs, R., Tuncer, B., Sariyildiz, S. (2002) Empowering individuals to design and build collaborative information spaces. *Proceedings, International Council for Research and Innovation in Building and Construction*, CIB w78, Aarhus school of Architecture.

Sun, M., Senaratne, S., Fleming, A., Motowa, I., Lin Yeoh, M. (2006) A change management toolkit for construction projects. *Architectural Engineering and Design Management*, 2(4): 261–271.

Tan, H.C., Anumba, C.J., Carrillo, P.M., Bouchlaghem, D., Kamara, J., Udeja, C. (2010) *Capture and Reuse of Project Knowledge in Construction*, Wiley-Blackwell: Oxford.

Tappeta, R.V., Renaud, J.E. (1997) Multiobjective collaborative optimization. *ASME Journal Mechanical Design*, 119: 403–411.

Technical Advisory Service for Images (TASI) (2006) *Metadata Standards and Interoperability*, Advice Paper, http://www.tasi.ac.uk.

Tevan, J., Dumais, S.T., Gutt, Z. (2008) Challenges for supporting faceted search in large, heterogeneous corpora like the web. *Proceedings from the Workshop on Human–Computer Interaction and Information Retrieval*, 13–15 March, Richmond, VA.

Tichy, N.M. (1983) *Managing Strategic Change*, Wiley: New York.

T-Mobile (2006) *Limited Adoption of Mobile Technology by the UK Construction Industry*, http://www.t-mobile.co.uk.

Tuckman, Bruce W. (1965) Developmental sequence in small groups. *Psychological Bulletin*, 63: 384–399. The article was reprinted in *Group Facilitation: A Research and Applications Journal*, 3, Spring 2001, and is available as a Word document: http://dennislearningcenter.osu.edu/references/GROUP%20DEV%20ARTICLE.doc.

Vakola, M., Rezgui, Y. (2000) Critique of existing business process reengineering methodologies: the development and implementation of a new methodology. *Business Process Management*, 6(3): 238–250.

Vakola, M., Wilson, I.E. (2002) The challenge of virtual organisation: critical success factors in dealing with constant change. *Proceedings of the European Conference on Information and Communication Technology Advances and Innovation in the Knowledge Society (eSM@RT 2002 in Collaboration with CISEMIC 2002)*, 18–21 November, Salford, UK.

Ward, M., Thorpe, T., Price, A., Wren, C. (2004) Implementation and control of wireless data collection on construction sites. *Electronic Journal of Information Technology in Construction (ITcon)*, 9: 297–311.

Warner Burke, W. (1994) *Organization Development: A Normative View*, 2nd edition, Addison-Wesley: Reading, MA.

Weiss, J.W. (2001) *Organizational Behaviour and Change: Managing Diversity, Cross-Cultural Dynamics, and Ethics*, South Western College Publishing: Cincinnati, OH.

Wilson, D.C., Rosenfeld, R.H. (1990) *Managing Organizations: Text, Readings and Cases*, McGraw-Hill: London.

WiMAX Forum (2004) Business case models for fixed broadband wireless access based on WiMAX technology and the 802.16 standard. Report, http://www.wimaxforum.org/technology/downloads/WiMAX-The_Business_Case-Rev3.pdf.

WiMAX Vision (n.d.) *Setbacks for WiMAX in Europe*, 3, http://www.wimax-vision.com/newt/l/wimaxvision/index.html.

Winograd, T. (1998) A language/action approach on the design of cooperative work. *Human Computer Interaction*, 3(1): 3–30.

Wireless LAN Alliance (WLANA) (n.d.) http://www.wlana.org/learning_center.html.

Wolstenholme, E.F. (1999) Qualitative vs quantitative modelling: the evolving balance. *Journal of the Operational Research Society*, 50(4): 422–428.

Woo, S., Lee, E., Sasada, T. (2001) The multiuser workspace as the medium for communications in collaborative design. *Automation in Construction*, 10(3): 303–308.

Woods, E. (2004) *Building a Corporate Taxonomy: Benefits and Challenges*, Ovum White Paper, http://www.ovum.com.

Yang, F. (2008) *Collaborative Optimisation in Building Design with a Pareto Based Genetic Algorithm*. PhD Thesis, Loughborough University.

Yassine, A., Braha, D. (2003) Complex concurrent engineering and the design structure matrix method. *Concurrent Engineering: Research and Applications*, 11: 165–175.

Yee, P., Swearingen, K., Li, K., Hearst, M. (2003) Faceted metadata for image search and browsing. *Proceedings of ACM CHI Montreal*.

Zolin, R., Hinds, P.J., Fruchter, R., Levitt, R.E. (2004) Interpersonal trust in cross-functional, geographically distributed work: a longitudinal study. *Information and Organisation*, 14: 1–26.

Index